Design of Prestressed Concrete

Series Editor: A. M. NEVILLE

Design of Prestressed Concrete

S. C. C. BATE and E. W. BENNETT

A HALSTED PRESS BOOK

JOHN WILEY & SONS
New York — Toronto

First published by
Surrey University Press
450 Edgware Road, London W2 1EG

First published 1976

Published in the U.S.A. and Canada by Halsted Press, a Division of
John Wiley & Sons, Inc., New York.

Library of Congress Catalog Card Number: 75-23527

Bate, S. C. C. and Bennett, E. W.
Design of Prestressed Concrete
New York   Halsted Press
1976 Jan.

0 470-05570-7

Printed in Great Britain by Robert MacLehose and Company Limited
Printers to the University of Glasgow

Foreword

This book forms part of a series covering various aspects of design and materials for the engineer.

The emphasis is on the word design as this is the crucial part of an engineer's activity, which distinguishes him from a scientist working in the related field. The engineer must be able to synthesize the information available for the purpose of creating on paper that which is to be built, i.e. designing. This activity requires knowledge of analysis of the whole as well as of the components of that which is being designed. It requires also knowledge of the codes and standards in force and what is considered good engineering practice. Finally, design requires engineering judgement, and judgement can come only with experience.

This is why a university course, even if design-oriented as at a few universities, does not produce a designer. The young graduate, however bright, when confronted with his first, and possibly not only first, design job simply lacks the background for his new task. This is where the books in the present series come in: they make it possible to approach the task of design in a reasonable manner.

The books cover mainly the field of structural design and include works on precast concrete, prestressed concrete and composite materials and formwork. These are the sort of design problems that the civil engineer is concerned with. The books explain relatively simply the background to the design problem and then in quite some detail, the main features of the design process. All this is fully illustrated by worked examples. Such an approach may seem old-fashioned to an enthusiast of pure analysis but example and precept are essential if modern design is to build upon the accumulated stores of successful design used in the past. Indeed, modern disciplines have borrowed our approach but re-named it 'case study'.

The books in the present series should thus prove of great value to the young engineer and also to his slightly senior colleague who is designing in an unfamiliar field. For these people the book is a 'must'. But it is also a wise investment (and in these inflationary days such is not easy to come by) for the undergraduate who appreciates the importance of design. With the aid of the book he can profit much more from his university or polytechnic course and enter employment much better prepared.

The authors of the books in the series are all specialists willing, in the best spirit of the engineering profession, to share their knowledge and experience. I am therefore confident that the series will be a success.

I can make these complimentary remarks as I have served only as general editor and not as author. There is therefore little doubt that the present series not only will fill a genuine need but will do so really well.

A. M. Neville
Leeds, 1975

Contents

List of Symbols

A_c	Area of concrete (may be shortened to A)
A_i	Area of *in-situ* concrete in a composite section
A_M	Area of the prestressing moment diagram
A_{ps}	Area of prestressing tendon (may be shortened to A_p)
A_s	Area of non-prestressed tension reinforcement
A_{pe}	Equivalent area of prestressed steel having strength equal to the combined strength of the prestressing tendons and non-prestressed tendon reinforcement
A_{sv}	Cross-sectional area of the parallel legs of a link or links within a distance s (may be shortened to A_v)
A_1	Area of precast part of a composite section
A_2	Area of composite section
a	Deflection
b	Breadth of section *or* Breadth of upper flange
b_e	Breadth of contact surface in a composite section
b_{inf}	Breadth of lower (inferior) flange in I or T section
b_w	Breadth of web
d	Effective depth of tension reinforcement
d_t	Depth from compression face to lowest bars or tendons around which links must pass
E_c	Modulus of elasticity of concrete
E_{ce}	Long-term (effective) modulus of elasticity of concrete
E_s	Modulus of elasticity of steel

e	Eccentricity of prestressing force with respect to centroid of concrete section
F	Load or force generally
F_k	Characteristic load
f	Direct stress or strength generally *or* Flexibility influence coefficient for unit reaction
f_{ci}	Cube strength of concrete at initial transfer of prestress
f_{cu}	Characteristic cube strength of concrete
f_{ct}	Direct tensile strength of concrete
f_{sup}	Prestress in concrete at top of section (superior)
f_{inf}	Prestress in concrete at bottom of section (inferior)
f_{cp}	Prestress in concrete at centroid of section
$f_{c\;adm}$	Allowable (admissible) compressive stress in concrete
$f_{tp\;adm}$	Allowable (admissible) tensile stress in concrete at initial transfer of prestress
f_{pb}	Tensile stress in tendons at ultimate moment of resistance of a beam
f_{pe}	Effective prestress in tendons
f_{pu}	Characteristic strength (ultimate stress) of tendons
f_s	Stress in non-prestressed tension reinforcement
f_y	Characteristic yield stress of reinforcement
f_{yv}	Characteristic yield stress of shear reinforcement
G_k	Characteristic concentrated dead load
g_k	Characteristic uniformly distributed dead load
h	Overall depth of section
h_f	Thickness of (upper) flange
h_{inf}	Thickness of lower (inferior) flange in I or T section
h_1	Overall depth of precast part of composite section
h_2	Overall depth of composite section
I_c	Second moment of area of uncracked concrete section (may be shortened to I)
I_1	Second moment of area of precast part of composite section
I_2	Second moment of area of composite section
i	Radius of gyration of concrete section
$\left.\begin{array}{c} K \\ k \end{array}\right\}$	Constants having dimensions
l	Effective span
M	Bending moment generally
M_d	Design moment (serviceability limit state)
M_{min}	Minimum moment acting on a prestressed section
M_0	Moment producing zero stress in concrete at level of tendon

M_u	Ultimate moment
M_F	Fixed-end moment
M_1	Moment acting on precast part of a composite section during construction
M_2	Moment acting on whole of composite section
N	Force normal to a section
N_{cu}	Force in concrete at ultimate moment of resistance of a beam
N_{pb}	Force in tendons at ultimate moment of resistance of a beam
P	Prestressing force
Q_k	Characteristic concentrated imposed load
q_k	Characteristic uniformly distributed imposed load
R	Redundant reaction
S	First moment of area of a concrete section to one side of a line parallel to the centroidal axis, about the axis
S_c	First moment of area of concrete section to one side of the interface, about the centroid of a composite section
s_v	Spacing of links (may be shortened to s)
u	Flexibility influence coefficient for external load
V	Shear force generally
V_c	Ultimate shear resistance of concrete
V_{co}	Ultimate shear resistance of concrete in a section uncracked in flexure
V_{cr}	Ultimate shear resistance of concrete in a section cracked in flexure
V_u	Ultimate shear force
V_{ud}	Design shear force for ultimate limit state
v	Shear stress, generally
v_c	Ultimate shear stress in concrete
v_h	Horizontal shear stress at interface of a composite section
v_u	Ultimate shear stress
v_1	Shear stress in precast part of a composite section
v_2	Shear stress in a composite section (additional to v_1)
W_k	Characteristic concentrated wind load
w_k	Characteristic uniformly distributed wind load
x	Linear co-ordinate *or* Depth of neutral axis
$\bar{x}$	Distance of centroid of area A_M from left-hand end of member
y	Vertical distance of a point from centroid of concrete section
y_{inf}	Distance of lowest (inferior) point from centroid of concrete section

y_{sup}	Distance of highest (superior) point from centroid of concrete section
y_o	Half depth of an anchorage block
y_{po}	Half depth of loaded area of an anchorage block
Z	Section modulus generally
Z_{inf}	Section modulus for lowest (inferior) point in concrete section
Z_{sup}	Section modulus for highest (superior) point in concrete section
Z_1	Section modulus for precast part of a composite section
Z_{2i}	Section modulus for composite section with reference to stress in *in-situ* concrete
Z_{2p}	Section modulus for composite section with reference to stress in precast concrete
z	Lever arm
α	Angle, ratio or dimensionless coefficient
α_e	Modular ratio of steel to concrete (may be shortened to α)
β	Angle, ratio or dimensionless coefficient
γ_m	Partial factor for materials
γ_f	Partial factor for forces
ϵ	Strain generally
ϵ_c	Strain in concrete
ϵ_{cs}	Limited shrinkage strain in concrete
ϵ_{cu}	Ultimate compressive strain in concrete
ϵ_s	Strain in steel
η	Reduction factor for loss of prestress
θ	Angle of inclination or deviation of tendons
μ	Coefficient of friction of tendon
ϕ	Dimensionless coefficient

Preface

This book has been written to cover the subject of prestressed concrete to the extent required by the final year engineering undergraduate and the structural engineer in the early years of practical experience. In it we have sought to give a concise exposition of the main principles, illustrated by worked examples relevant to design practice. The background of the book is the British Standard Code of Practice for the Structural Use of Concrete (CP 110: 1972), but we also refer to the American Code AC1 318:71 which may be helpful to transatlantic readers.

In the interests of completeness we have commenced with a short account of the limit state approach to design before going on to explain the respective methods of analysis of prestressed concrete for the serviceability and ultimate limit states. This section includes a number of tables for reference and to reduce the length of some of the routine calculations. There follows the description of a method of selecting and dimensioning prestressed members with pretensioned or post-tensioned tendons and the succeeding examples introduce some of the standard sections which have now been fairly widely adopted. A chapter is devoted to composite construction which is one of the most extensive current applications of prestressed concrete, particularly in highway bridge construction, and the book concludes with an introduction to the methods of analysis and design of statically indeterminate prestressed structures.

We gratefully acknowledge the cheerful and willing help received from Miss Christine Morris, Mrs. Gertrud Uddin and Mrs. Diane Uezzell in the preparation and correction of the typescript.

1

Limit State Design of Prestressed Concrete

1.1 INTRODUCTION

Prestressing is a technique whereby the performance of a structure is improved by the introduction of permanent stress (prestress) so as to cancel some of the stress produced by the dead and imposed loading. Concrete, with its inherent weakness in tension, is a particularly suitable material for prestressing since a compressive prestress can be used to reduce the tensile stress to an acceptably low value or, if desired, to eliminate it altogether.

Prestressed concrete structures exhibit greatly improved qualities under service load conditions on account of the absence of the cracking which characterizes conventional reinforced concrete in tension and flexure. The steel is afforded greater protection, there is increased resilience and because the whole of the concrete cross-section is effective, lighter members can be designed. A reduction in size is also possible when there is a large dead load, most of the stress due to which can be balanced by prestress.

Permanent forces are necessary to produce prestress; these may be set up by several different methods, the commonest of which is the introduction of tendons, consisting of tensioned high-strength steel wires, stranded cables or bars. The tendons pass through the structural members to which they are anchored at the ends, thereby applying a compressive reaction to the concrete. High strength steel at a high initial stress is essential to offset the loss of stress due to inelastic contraction of the concrete due to creep and shrinkage.

Prestressed concrete became a practical possibility in the 1930's when the phenomena of creep and shrinkage began to be understood, and it was realized that the earlier failures with mild steel tendons could be avoided by the use of high tensile steel. After the war, development in Europe, and a little later in America, was rapid and the first attempt in the United Kingdom to codify a design procedure was made in 1951 when the Institution of Structural Engineers published its First Report on Prestressed Concrete ref. (1.1). This Report formed the basis for the preparation of the British Standard Code of Practice CP 115, entitled The Structural Use of Prestressed Concrete in Buildings, ref. (1.2), which was published in 1959. Although still in current use, this is now being supervised by the unified Code CP 110 ref. (1.3) issued by the British Standards Institution in 1972. The present text will refer mainly to the latter Code, with some additional references to the current American Code ACI 318–71, Building Code Requirements for Reinforced Concrete (ref. (1.4)).

The main difference in concept between design to CP 115 and that to CP 110 is the introduction of the limit state approach in the latter Code. The changes in procedure between the two Codes were however small and produced none of the difficulties experienced with changes in procedure required for reinforced concrete in adopting limit state design. The reason for this was that by its nature, prestressed concrete required separate consideration of its behaviour under normal conditions of loading and service from its performance under exceptional conditions such as overloading. Thus design dealt with both working loads and ultimate loads, and hence CP 115 anticipated the needs to consider serviceability and ultimate limit states, which were later to become an inherent part of limit state design philosophy.

A fundamental aspect of the application of limit state design is that the designer should make an objective assessment of the problems in the design of each structure with which he is concerned, dealing in the initial planning stages with each structure separately and avoiding the use of automatic and routine design methods until it has been established that they will give the degree of safety and serviceability required. Structural failures are not infrequently the result of the successive applications of design procedures to developing forms of construction to which they become increasingly unsuited. The outline description of limit state design given in this chapter deals with its application to prestressed concrete only. It aims at providing a no more than adequate background to the intentions given in the design methods advocated so that the designer can use his own judgement on their suitability for application to any design requirements even when these are out of the ordinary.

Early structural Codes of Practice provided margins of safety by adopting permissible stresses for the materials defined as fractions of their strengths. Structures were usually envisaged as simple combinations of structural elements and the analysis of the stresses in the materials were almost always based on elastic theory. Loads were precisely defined with the assumption that they would not be exceeded in service and no account was taken of the variations in the strengths of the materials. In limit state design, the designer is required to examine all the conditions (limit states) that might result in the structure being found unfit for its intended purpose. He is then further required to provide a design which ensures that the risk that a limit state is reached is acceptably remote taking account of the likelihood of material damage and injury or loss of life on the one hand and the structural cost of avoiding the limit state on the other. Furthermore since this approach requires a more realistic appraisal of the design, it is more readily able to incorporate new research data on loads and hazards that must be sustained, on the variability of materials and workmanship and on methods of analysis. Although this may lead to greater design complexity, this sophistication has become necessary as a result of demanding more from the performance of structures and of using materials of greater strength and lower ductility.

The Code of Practice CP 110 states that all limit states should be considered but gives specific requirements for serviceability and ultimate limit states, conditions such as fire and durability being dealt with by arbitary provisions. It sets out basic functional requirements for these limit states, the safety factors that should be used for enhancing loads and reducing strengths for use in the design calculations to provide the margins of safety needed for each limit state, and certain *ad hoc* rules for safeguarding structures against a disproportionate amount of failure in the event of accidental damage.

1.2 DESIGN LOADS

The design load for each limit state F_d is

$$F_d = \gamma_f F_k$$

where F_k is the characteristic load which is independent of the limit state considered and is given in CP 3 Chapter V, Parts 1 and 2 for imposed and wind loads respectively. It is the load which is seldom likely to be exceeded in service and, for example for wind loading, is unlikely to occur more than once during the life of the structure.

γ_f is the partial safety factor for loads and load effects, with a value depending on:

(a) the importance of avoiding the limit state being considered;
(b) the possibility of experiencing loads in excess of those considered likely in CP 3 Chapter V, Parts 1 and 2, ref. (1.5);
(c) the likelihood of various loads reaching high values at the same time;
(d) inaccuracies in assessing loading effects and the effect of inaccuracy in construction on loads.

Values assigned to γ_f, the partial safety factor for loads and load effects are given in Table 1.1 for both serviceability and ultimate limit states.

Table 1.1

Combination of loads		Design load F_d for limit state incorporating partial safety factors γ_f
Serviceability limit states		
dead and imposed loads		$1 \cdot 0 G_k + 1 \cdot 0 Q_k$
	or	$1 \cdot 0 G_k$
dead and wind loads		$1 \cdot 0 G_k + 1 \cdot 0 W_k$
dead, imposed and wind loads		$1 \cdot 0 G_k + 0 \cdot 8 Q_k + 0 \cdot 8 W_k$
Ultimate limit states		
dead and imposed loads		$1 \cdot 4 G_k + 1 \cdot 6 Q_k$
	or	$1 \cdot 0 G_k$
dead and wind loads		$0 \cdot 9 G_k + 1 \cdot 4 W_k$
	or	$1 \cdot 4 G_k + 1 \cdot 4 W_k$
dead, imposed and wind loads		$1 \cdot 2 (G_k + Q_k + W_k)$

Where G_k = characteristic dead load

    Q_k = characteristic imposed load

    W_k = characteristic wind load

Where alternatives are given, the more severe combinations of loads should be considered. In a cantilevered structure for example subjected to dead and wind loads at the ultimate limit state, the factor for that portion of the dead load contributing to stability would be $0 \cdot 9$ whilst the factor for the portion of dead load detracting from stability would be $1 \cdot 4$. Similarly in the design of continuous beams when alternate spans are assumed to be loaded, the loaded spans are assumed to support a load of $1 \cdot 4 G_k + 1 \cdot 6 Q_k$ whilst the unloaded spans are assumed to be carrying $1 \cdot 0 G_k$.

A similar set of factors with slightly different values is given for the determination of required strength in ACI 318–71 9.3.

1.3 DESIGN STRENGTHS

The strengths of materials used in design for each limit state i.e. the design strengths f_d are given by:

$$f_d = \frac{f_k}{\gamma_m}$$

where f_k is the characteristic strength of the material and is independent of the limit state being considered. It is $f_m (1 - 1 \cdot 64 \sigma_f)$ where f_m is the mean strength and σ_f is the standard deviation of strength as determined by tests.

The coefficient 1·64 ensures that not more than 5% of test results will fall below f_k.

γ_m is the partial safety factor for materials with a value depending on the importance of the limit state being considered, the material to which it applies and differences between the strengths of the materials as tested and those when incorporated in construction during the service life of the structure. The design strengths for each limit state are obtained using the values for the partial safety factors γ_m for the materials given in Table 1.2.

The American Code specifies that the calculated strength of a member or section should be multiplied by a single reduction factor, values for which are given in ACI 318–71, 9.2.

Table 1.2

States considered and material	Partial safety factor γ_m
Serviceability limit states	
steel	1·0
concrete—deflection	1·0
—cracking	1·3
Ultimate limit states	
steel	1·15
concrete	1·5

1.4 CRITERIA FOR SERVICEABILITY

Whilst the criteria which define the ultimate limit states are easily recognized since they are associated with partial or complete collapse of the structure, those conditions which represent the serviceability limit states can only be stated much less specifically. The limits given in the Code deal with deflection and cracking; for deflection, they are as follows:

Final deflection of beams and slabs $\not> \dfrac{\text{span}}{250}$

Where partitions or finishes might be affected the final deflection $\not> \dfrac{\text{span}}{350}$ or

$\not> 20 \text{ mm}$

If appropriate for lightly loaded prestressed concrete units where finishes are to be applied, upward deflection $\not> \dfrac{\text{span}}{300}$

These values have been selected from experience and are not the consequence of scientific investigation; some latitude in interpretation may therefore be appropriate. No limit has been set for the horizontal deflection of frames although a figure of height/500 might be considered appropriate.

The Code CP 110 is much less specific in setting acceptable limits for cracking in prestressed concrete construction. Structures are divided into three classes but no guidance is given on what circumstances would govern the choice of category in design. The requirements for these classes are as follows:

Class 1 *structures*—no tensile stresses are permitted in the concrete and hence a margin against cracking is aimed at.

Class 2 *structures*—tensile stresses but no cracks are permitted and hence stresses in the concrete are limited to values less than the tensile strength of the concrete.

Class 3 *structures*—cracks are permitted provided that their maximum surface width does not exceed 0·2 mm for normal conditions or 0·1 mm when conditions of exposure are particularly aggressive. Nominal tensile stresses are recommended in the Code for use in calculations, which have been shown by experiment to conform with the limitation on crack width.

1.5 APPROACH TO DESIGN

The requirement in the Code that all relevant limit states should be considered and in particular that ultimate and serviceability limit states should be checked is no more onerous for prestressed concrete than the requirements of the earlier Code, CP 115. Experience with particular classes of structure will indicate the most suitable sequence of the calculations. For Class 1 structures, the limitation of no bending tensile stress in the concrete may well govern design, whereas for Class 3 structures, the ultimate limit state is likely to govern and should be treated first in design as in the treatment of reinforced concrete according to CP 110.

References

1.1 'First report on prestressed concrete', Institution of Structural Engineers, London, 1951.

1.2 BRITISH STANDARDS INSTITUTION, The Structural Use of Prestressed Concrete in Buildings, British Standard Code of Practice, CP 115: 1959.

1.3 BRITISH STANDARDS INSTITUTION, The Structural Use of Concrete, British Standard Code of Practice, CP 110: 1972.

1.4 AMERICAN CONCRETE INSTITUTE, Building Code Requirements for Reinforced Concrete (ACI 318-71), 1971.

1.5 BRITISH STANDARDS INSTITUTION, Basic Data for the Design of Buildings— Loading, British Standard Code of Practice, CP 3 Chapter V, Part 1: 1967 and Part 2: 1972.

2

Analysis for Serviceability Limit States

2.1 BASIS FOR ANALYSIS

The analysis of prestressed concrete structures for serviceability limit states is based on the assumption that deformation of the materials is elastic, the effects of non-linear deformation of the concrete due to creep and shrinkage and of the steel due to relaxation being dealt with by minor modifications to the elastic theory. Thus the moments and forces acting on a prestressed concrete structure and the resulting stresses are determined by established methods of elastic analysis for these limit states. In the analysis of structures and in the calculation of stresses, it is normally sufficient to use the geometric properties of the concrete cross-section neglecting the effect of the steel tendons.

The design loads for the serviceability limit states are derived from the factors given in Table 1.1. With the exception of the combined effects of dead, imposed and wind load, γ_f is taken as $1 \cdot 0$. For most combinations of loading therefore, the design loads for serviceability have their characteristic values.

The majority of prestressing systems fall into two main categories. Members with pretensioned tendons are those in which the steel is tensioned before placing the concrete, as opposed to post-tensioned tendons which are stressed after the concrete has been placed and has hardened. The term pretensioned or post-tensioned refers to the tendons and not to the member itself, which is more correctly described as being prestressed with pretensioned (or post-tensioned) tendons. Most prestressed concrete members stressed by pretensioning are incorporated in construction as simply supported beams, and the extremes of loading which need to be examined at the serviceability limit state for cracking and local damage are:

(a) at transfer,
(b) during handling, particularly for long or slender members,
(c) under full design load after allowance for all losses of prestress.

Such members are often incorporated in composite construction and may sometimes be made continuous by reinforcing the *in situ* concrete to develop moments of resistance over the supports. In these circumstances, the situations needing to be checked are modified and will be discussed further in Chapters 6 and 7.

Post-tensioning offers greater flexibility in design not only because the position of the tendons can be varied along the beam but also because the tendons can be stressed at appropriate stages as construction proceeds.

Checking for conformity with the serviceability requirements for deflection will not be made at the stages outlined in the previous paragraph. For beams with pretensioned tendons, limitation on upward deflection to 1/300 of the span applies only to lightly loaded beams of long span such as in roof construction. The need for uniformity of upward deflection is in composite construction such as floors and is obtained by supervision and control of production rather than by calculation. Downward deflection will not always need to be checked as discussed later. When this is necessary however, the check should be made for the imposed loads that are likely to be permanently applied and not for the full design load, allowance being made for the effects of creep and shrinkage of the concrete and other causes of loss.

The partial safety factors for the materials given in Table 1.2 for serviceability limit states define γ_m as 1·0 except for the design strength of concrete to be used in assessing resistance to cracking where it has the value of 1·3. The partial safety factors have been incorporated in the design approach adopted in CP 110. In the Code, the design of sections for the avoidance of local damage requires the consideration of the stresses in the concrete resulting from the various possible combinations of prestress, adjusted as necessary for the appropriate losses of prestress, and of the external forces and moments induced by the loads; these stresses should not exceed the values allowed in CP 110 for the corresponding age and quality of the concrete. In choosing the allowable values given in CP 110, the values of the partial safety factors for the materials have already been taken into account. Thus the levels of stress allowed provide for compliance with the limit state requirements for local damage to the concrete, and in particular the allowable tensile stresses ensure that the recommendations for Class 1, Class 2 and Class 3 structures are met with respect to cracking.

2.2 STRESSES IN THE CONCRETE

A beam subjected to the prestress alone is essentially a prismatic member subjected to eccentric loading, and hence the stresses in the concrete are as follows (see Figure 2.1)

$$f_{sup} = \frac{Pey_{sup}}{I} - \frac{P}{A} \tag{2.1}$$

$$= \frac{P}{A}\left(\frac{ey_{sup}}{i^2} - 1\right) \tag{2.2}$$

where, f_{sup} = tensile stress at top of section
A = the area of the concrete section
I = the second moment of area of the concrete section, and
i^2 = $\dfrac{I}{A}$

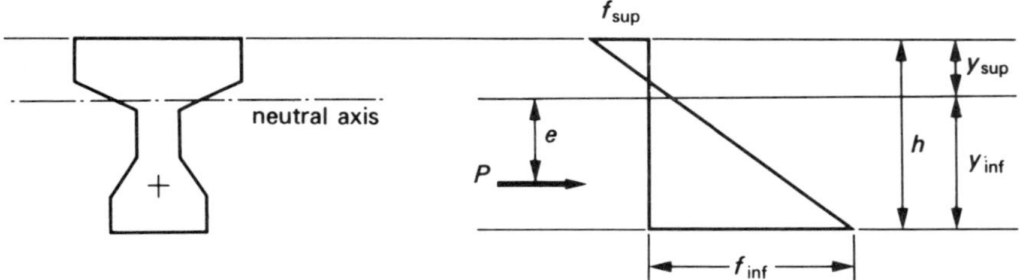

Figure 2.1

Similarly, the compressive stress at the bottom of the section is given by

$$f_{inf} = \frac{P}{A}\left(\frac{ey_{inf}}{i^2} + 1\right) \tag{2.3}$$

The stresses in the concrete due to an external moment M producing compression at the top and an external compressive force N are as shown in Figure 2.2, namely,
at the upper surface a compressive stress of

$$\frac{M}{Z_{sup}} + \frac{N}{A} \tag{2.4}$$

at the lower surface a tensile stress of

$$\frac{M}{Z_{inf}} - \frac{N}{A} \tag{2.5}$$

where, Z_{sup} and Z_{inf} are respectively the section moduli for the upper and lower sections respectively.

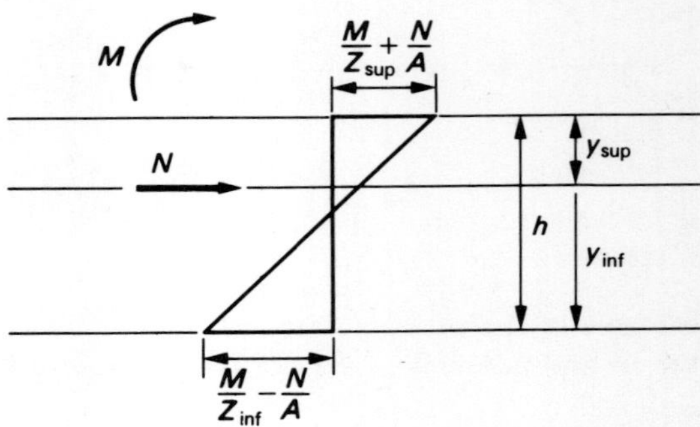

Figure 2.2 Concrete stresses due to external moment M and force N

For most prestressed concrete construction, there is no longitudinal force N and the second term in (2.4) and (2.5) then disappears.

The stresses in the concrete at any stage may then be assessed from the sum of the values obtained from (2.2) and (2.4) and from (2.3) and (2.5) taking account of the external moments and forces then acting and the appropriate level of prestress.

Examples

The steel tendons in a rectangular prestressed concrete beam, which is 150 mm wide and 600 mm deep, are located at a mean depth of 390 mm. If the force in the tendons is 1200 kN, calculate the stresses in the concrete ignoring the selfweight of the beam.

Referring to expressions (2.2) and (2.3)

$$A = 250 \times 600 = 150\ 000\ \text{mm}^2$$

$$i^2 = \frac{I}{A} = \frac{d^2}{12} = 30\ 000\ \text{mm}^2$$

$$y_{sup} = y_{inf} = 300\ \text{mm}$$

$$e = 390 - 300 = 90\ \text{mm}$$

Whence,

$$f_{sup} = \frac{1200 \times 1000}{150\ 000} \left(\frac{90 \times 300}{30\ 000} - 1 \right) = -0 \cdot 8\ \text{N/mm}^2$$

Thus the stress in the concrete at the top surface is compressive.

Similarly,

$$f_{\text{inf}} = \frac{1200 \times 1000}{150\ 000}\left(\frac{90 \times 300}{30\ 000} + 1\right) = 15\cdot2\ \text{N/mm}^2$$

If this beam is supported on a span of 10 m, calculate the uniform load that it could support without developing bending tensile stresses.

Referring to expressions (2.4) and (2.5), the tensile stress due to the moment at the lower surface should not exceed $15\cdot2$ N/mm² if tension is to be avoided.

$$Z_{\text{inf}} = \frac{250 \times 600^2}{6} = 15 \times 10^6\ \text{mm}^3$$

from (2.5) $M = Z_{\text{inf}} \times 15\cdot20 = 228$ kNm.

Whence the uniform load on a span of 10 m is

$$\frac{228 \times 8}{10} = 182\ \text{kN}$$

i.e. about five times the weight of the beam.

2.3 LOSSES OF PRESTRESS

In most prestressed construction, the prestress in the concrete is imposed by steel tendons, which are anchored to the concrete and are usually, but not always, located within the section. With both pretensioning and post-tensioning, a reduction in the stress in the tendons occurs at transfer and subsequently there is a further long-term reduction which leads to a substantial loss of prestress and has an important influence on design for serviceability.

In the pretensioning systems, losses of prestress result from elastic contraction of the concrete when the prestress is transferred, and later as a result of creep and shrinkage of the concrete and relaxation of the steel itself. If steam curing is used in the production of units, the losses may be further increased. With post-tensioning, losses of prestress also result from elastic contraction, creep and shrinkage of the concrete and from relaxation of the steel, but overall they tend to be smaller than with pretensioning partly because the concrete is usually more mature at the time of transfer of the prestress to the concrete. In addition however, other losses may occur in post-tensioning due to friction in the ducts during stressing and to movement in the grips of wedge-type anchorages during anchoring; this latter cause of loss is only important for short tendons and may be avoided by using bars with a nut for anchoring.

B

2.4 ELASTIC CONTRACTION OF THE CONCRETE

The reduction of stress in the tendons resulting from the shortening of the concrete at transfer is determined from the conditions of stress and the respective moduli of elasticity of the materials. For pretensioning, the stress in all the tendons is transferred to the concrete at the same time and therefore the stress f_{pp} in the tendons after transfer is given by

$$f_{pp} = f_{pi} - \alpha f_{cp} \tag{2.6}$$

where, f_{pi} = the stress in the tendons before transfer, which is usually taken as the stress due to the initial jacking although strictly any relaxation of stress in the steel between initial stressing and transfer should be taken into account.

f_{cp} = the stress in the concrete at the level of the tendons after transfer.

α = the ratio of the moduli of steel and concrete, E_s/E_c.

The moduli of elasticity for concretes of different strength as recommended in CP 110:1972 are given in Table 2.1; Table 2.2 gives the moduli of elasticity of the different types of tendon also derived from CP 110.

The extent of the loss of prestress that may occur at transfer with pretensioning is shown in Table 2.1 for a range of steels and concretes. The percentage loss of stress in the tendons is obtained by multiplying the appropriate figure from one of the last two columns by f_{cp}.

For post-tensioning, there may be a number of tendons in a single member which are not stressed simultaneously. As each tendon is stressed therefore, it causes a small contraction in the concrete member and so reduces the stress in those tendons which have already been stressed and anchored. The loss of stress resulting from this elastic contraction of the concrete varies from about half that given by (2.6) above for beams with a large number of tendons to zero in beams with a single tendon. In general therefore, the loss of stress for post-tensioning may be taken as approximately half that in Table 2.1.

2.5 RELAXATION OF THE STEEL IN THE TENDONS

The steels used for tendons are high tensile materials which are stressed to a high proportion of their tensile strength to minimize the losses of prestress due to deformation of the concrete. It is recommended in CP 110 that the jacking force should not normally exceed 70% of the characteristic strength of the tendon but, provided additional consideration is given to safety, the characteristics of the steel, and to assessing friction losses, it may be increased to 80%.

The types of steel, which are available and are covered by British Standards, are listed in Table 2.2. A characteristic of all these steels is that they relax when

Table 2.1 *Elastic loss of prestress at transfer for pre-tensioning*

Required strength of concrete at transfer/N/mm^2	Modulus of elasticity of concrete at transfer/kN/mm^2	Characteristic strength of tendon/N/mm^2	Modulus of elasticity of the steel/kN/mm^2	Percentage loss of stress in the tendon for each 1 N/mm^2 of stress in the concrete at the steel centroid	
				$f_{pi}/f_{pu} = 0.7$	$f_{pi}/f_{pu} = 0.8$
30	28	1000*	175	0·89	—
		1500	175	0·60	0·52
		1500	200	0·68	0·60
		1750	200	0·58	0·51
40	31	1000*	175	0·81	—
		1500	175	0·54	0·47
		1500	200	0·61	0·54
		1750	200	0·53	0·46
50	34	1000*	175	0·74	—
		1500	175	0·49	0·43
		1500	200	0·56	0·49
		1750	200	0·48	0·42
60	36	1000*	175	0·69	—
		1500	175	0·46	0·41
		1500	200	0·53	0·46
		1750	200	0·45	0·40

* Tendons with a characteristic strength of 1000 N/mm^2 are bars which are seldom used for pretensioning.

Table 2.2 *Loss of prestress due to relaxation of steel*

Type of tendon	Range of characteristic strength/N/mm²	Modulus of elasticity/kN/mm²	Percentage loss of prestress	
			$f_{pi}/f_{pu} = 0.7$	$f_{pi}/f_{pu} = 0.8$
cold drawn steel wire to BS 2691:				
i prestraightened—normal relaxation	1570–1720	200	5	8·5
ii prestraightened—low relaxation	1570–1720	200	2	3
iii wire in mill coils	1550–1720	200	8*	10*
seven wire strand to BS 3617:				
i normal relaxation	1640–1820	200	7	12
ii low relaxation	1640–1820	200	2	3
nineteen wire strand to BS 4757:				
i as spun	1480–1540	175	9	14
ii normal relaxation	1760	175	7	12
iii low relaxation	1760	175	2·5	3·5
cold worked high tensile alloy steel bars to bars to BS 4486	995–1030	175	4	—

* Note: These figures are not specified in the British Standard; the other values are maxima to be satisfied by test.

Table 2.3 *Loss of prestress due to creep of concrete*

| Required strength of concrete at transfer/N/mm^2 | Characteristic strength of tendon/N/mm^2 | Modulus of elasticity of the steel/kN/mm^2 | Percentage loss of stress in the tendon for each 1 N/mm^2 of stress in the concrete at the steel centroid | | | |
| | | | Pretensioning | | Post-tensioning | |
			$f_{pi}/f_{pu} = 0.7$	$f_{pi}/f_{pu} = 0.8$	$f_{pi}/f_{pu} = 0.7$	$f_{pi}/f_{pu} = 0.8$
30	1000	175	—	—	1·20	—
	1500	175	1·07	0·93	0·80	0·70
	1500	200	1·22	1·07	0·91	0·80
	1750	200	1·04	0·91	0·78	0·69
40	1000	175	—	—	0·90	—
and	1500	175	0·80	0·70	0·60	0·52
over	1500	200	0·91	0·80	0·69	0·60
	1750	200	0·78	0·68	0·59	0·51

Notes: 1. If the maximum stress in the section at transfer exceeds one-third of the cube strength of the concrete these values should be increased by a factor which has a value of 1·25 when the stress is half the cube strength.
2. The calculations are based on values for creep per 1 N/mm^2 of 48×10^{-6} for prestressing and 36×10^{-6} for post-tensioning. For concrete with a cube strength of less than 40 N/mm^2 at transfer, the values have been increased in inverse proportion.

Table 2.4 *Loss of prestress due to shrinkage of concrete*

System for stressing	*Environment*	*Characteristic strength of tendon*/N/mm²	*Modulus of elasticity of steel*/kN/mm²	*Percentage loss of prestress*	
				$f_{pi}/f_{pu} = 0.7$	$f_{pi}/f_{pu} = 0.8$
pretensioning	humid (90% RH)	1500	175	2	1
		1500	200	2	2
		1750	200	2	1
	normal (70% RH)	1500	175	5	4
		1500	200	6	5
		1750	200	5	4
post-tensioning	humid (90% RH)	1000	175	2	—
		1500	175	1	1
		1500	200	1	1
		1750	200	1	1
	normal (70% RH)	1000	175	5	—
		1500	175	3	3
		1500	200	4	3
		1750	200	3	3

Note: The calculations are based on values for shrinkage of 300×10^{-6} for normal and 100×10^{-6} for humid conditions for pretensioning, and of 200×10^{-6} for normal and 70×10^{-6} for humid conditions for post-tensioning.

strained at constant length under high stress. This reduction of stress with time is a further source of loss of prestress. The values given in Table 2.2 for this loss for two different levels of initial stress are in each case the relaxation after 1000 hours at constant strain as specified in the appropriate British Standard. These figures are adopted for design since, although relaxation continues indefinitely at a reducing rate, the conditions in the prestressed member are those of reducing strain resulting from creep and shrinkage of the concrete.

2.6 CREEP AND SHRINKAGE OF THE CONCRETE

When concrete is subjected to load, the initial elastic deformation is followed by a further and usually greater amount of deformation which takes place gradually with time. Part of this deformation is dependent on the level of applied stress and is termed creep, whilst the other part is independent of the applied stress and is described as shrinkage. Both creep and shrinkage of the concrete contribute substantially to the losses of prestress. For convenience, it is assumed that creep is proportional to stress for which there is some experimental support for stresses up to about one-third of the cube strength; at higher stresses creep becomes disproportionately greater.

CP 110 gives design values for both creep and shrinkage, which allow for the greater maturity of the concrete for post-tensioning as compared with pretensioning, and which differentiate between humid and normal conditions for shrinkage. These values are given in the footnotes to Tables 2.3 and 2.4 and are used to derive the tables. These illustrate the percentage loss of stress in the tendons for different combinations of materials and have been obtained from the product of the contraction induced in the steel and its modulus of elasticity; the percentage loss due to creep is determined by multiplying the values given by f_{cp}, the stress in the concrete at the centroid of the steel.

2.7 TOTAL LOSS OF PRESTRESS DUE TO DEFORMATION OF THE MATERIALS

Since the prediction of the behaviour of the materials, particularly the concrete, is open to considerable error at the design stage, it is sufficiently accurate for the assessment of the prestress conditions to add the losses of stress together to obtain the overall loss due to the deformation of the materials.

Examples

(a) *If the strength of concrete at transfer is to be 40 N/mm² for a normally exposed beam with pretensioned low-relaxation 7-wire strand stressed to 80% of its characteristic strength of 1750 N/mm², and the stress in the concrete at the steel centroid is 10 N/mm² after transfer, estimate the total loss of prestress.*

Loss due to

elastic contraction	$= 0.46 \times 10$	$= 5\%$	from Table 2.1
relaxation of steel		$= 3\%$	from Table 2.2
creep of concrete	$= 0.68 \times 10$	$= 7\%$	from Table 2.3
shrinkage		$= 4\%$	from table 2.4

$$\text{total} \quad 19\%$$

(b) *If the same beam were to be prestressed when the cube strength is* 30 N/mm^2 *using drawn wire from mill coils stressed initially to* 70% *of the characteristic strength of* 1550 N/mm^2, *again estimate the total loss of prestress.*

Loss due to

elastic contraction	$= 0.68 \times 10$	$= 7\%$
relaxation of steel		$= 8\%$
creep of concrete	$= 1.22 \times 10 \times 1.25^*$	$= 15\%$
shrinkage		$= 6\%$

$$\text{total} \quad 36\%$$

Although these estimates are approximate, they nevertheless show that changes in materials and procedures may make substantial changes to the losses of prestress obtained. In Chapter 4 and the following Chapters, the loss ratio η is introduced, which is the ratio of the final prestress in the concrete to the prestress first applied. This ratio does not include therefore, the effect of elastic contraction and, for the examples given, the values of η would be $(1.00 - 0.05 - 0.03 - 0.07 - 0.04)/(1.00 - 0.05) = 0.85$ for (a) and similarly 0.69 for (b). Some relaxation of the tendons occurs before transfer, particularly with pretensioning, and strictly should be taken into account with the loss due to elastic contraction, but for practical purposes it may be assumed to occur later. Where reliable experimental data are available on the actual deformation characteristics of the materials to be used in the construction, these should be used in design.

2.8 FRICTION

An advantage of prestressing by post-tensioning is that the tendons may be positioned in such a way that the most favourable distribution of stresses can be obtained at different sections. As a consequence, post-tensioned tendons will normally be bent up and down as circumstances require and friction will

---

* The factor 1.25 is introduced since the maximum stress in the concrete after transfer is about half the cube strength.

therefore develop during stressing wherever changes in direction are imposed. The tendons will be located either within ducts in the concrete or by external guides. Since some inaccuracy in positioning the ducts or other forms of restraint is inevitable on site, friction is experienced even with straight tendons and the level of supervision and workmanship recommended in CP 110. If supervision and workmanship are not good and grout is allowed to penetrate the ducts, for example, full stressing of tendons to design requirements may not be possible and in extreme conditions tendons may be fractured during stressing.

In design, the loss due to friction is taken as comprising two components, one due to accidental deviation or 'wobble' of the tendon from its proper position and the other due to the deliberate change of direction in the design. In general, the loss expressed as a percentage is given by

$$(1 - e^{-(Kx + \mu\theta)}) \times 100 \tag{2.7}$$

where, K = a constant related to the unintended 'wobble' which generally has a value of 33×10^{-4} per metre but may be taken as 17×10^{-4} per metre where well fixed rigid sheaths or duct formers are used.

x = the length of the tendon in metres along which the loss is being calculated.

θ = the angular deviation of the tendon in the length x.

μ = the coefficient of friction which has the following values for beams:

    0·55 for steel moving on concrete,
    0·30 for steel moving on steel,
    0·25 for steel moving on lead,

and for circular construction:

    0·45 for steel moving on concrete,
    0·25 for steel moving on steel bearers fixed to the concrete,
    0·1 for steel moving on steel rollers.

Other values for the coefficient of friction may be used when it has been shown by development testing that they can be obtained consistently in practice. When the value of $(Kx + \mu\theta)$ is less than 0·2 the percentage loss is given with sufficient accuracy as

$$(Kx + \mu\theta) \times 100 \tag{2.10}$$

that is for losses up to 20%.

In beams, bond between the tendons and the concrete is normally established after stressing by grouting the ducts with cement grout. If lubricants are used to reduce friction, their effect on the bond achieved later should be considered in the design.

Friction may also develop during stressing where the tendons pass through the anchorage, and information on the magnitude of this loss should be supplied by the manufacturer. Many forms of anchorage rely on wedge-grips to hold the ends of the tendons after stressing; in developing their grip some loss of extension of the tendon results, which is substantial for short members. Allowance for this movement should be made in the jacking operations, during which it should be measured, but in deciding the initial stress after transfer at the design stage, allowance should also be made for this movement from information supplied by the maker.

2.9 TRANSMISSION LENGTH OF TENDONS

The stress in each pretensioned tendon increases from zero at the end to the transfer value at the end of the transmission length. This increase is not linear but occurs at a greater rate near the end so that 80% of the transfer stress may be developed in about 70% of the transmission length. The effect of creep of the concrete is to increase the transmission length slightly over a period of time, and the sudden transfer of the prestress, for example by cutting the wires while still under tension, will result in an increased transmission length.

It is generally sufficient for the designer to know a safe maximum value of the transmission length, and as there has been found to be considerable variation it is best for this to be based on measurements at the particular site or factory. The main factors affecting the transmission length are the form and surface condition of the tendons and the quality of the concrete. The length will be shortened for tendons of small diameter, or where the surface is roughened by moderate rusting, or the tendon is deformed. Strands are very effective in transmitting prestress and wires of up to 7 mm diameter may be used, the larger sizes being indented, or preferably crimped to improve the bond. The transmission length tends to be less with well compacted high strength concrete, and may be greater for tendons at the top of a beam owing to poorer compaction or bleeding.

The transmission lengths given as general recommendations in CP 110 are 100 diameters for plain or indented wire with a small offset crimp (e.g. 0·3 mm offset, 40 mm pitch) and 65 diameters for wire with a considerable crimp (e.g. 1·0 mm offset 40 mm pitch). These values apply to fully compacted concrete with a cube strength of at least 35 N/mm^2 at transfer. The transmission length for strand is not proportional to the diameter, and values are quoted for three different sizes. When pretensioned tendons are located in a relatively small area of concrete, bursting stresses in the concrete will develop along the transmission length. For strand, this region should be reinforced with closed links; the amount of this reinforcement is probably best determined by trial.

2.10 ALLOWABLE STRESSES

Having considered the magnitude of the losses of prestress that are likely to occur, it is now necessary to define the stresses for use in design to ensure that an unacceptable level of cracking or damage to the concrete is avoided when the prestress is applied and later under the action of the dead and imposed loads. The values of these allowable stresses are so chosen that the requirements of the serviceability limit state are met with respect to damage but at the same time these allowable values also help to ensure that deflections are not excessive.

In the Code of Practice CP 110 and in ACI 318–71, 18.4 the allowable stresses are given only for members subjected to conditions of flexure and direct load. Shear is dealt with only when considering the ultimate limit state, since the web will normally be uncracked and the stresses in the concrete low under service load conditions.

The three classes of prestressed concrete structure have already been referred to in section 1.4 with the corresponding limitations on the amount of cracking that is acceptable, although CP 110 does not indicate the appropriate uses for each class. It may be observed however that this classification represents a gradual transition from structures which are expected to be completely free from cracking in service on the one hand, as represented by Class 1 structures, to those which are expected to be cracked in service as represented by reinforced concrete construction on the other; construction in Classes 2 or 3 is intermediate. Clearly if the environment is particularly aggressive or if water retention is required, there are merits in adopting the stress limitations for Class 1 structures. Generally, the use of lower tensile stresses with the consequent greater freedom from cracking is likely to lead to smaller deflections in service and hence if deformation has to be restricted Class 1 or 2 construction may be appropriate. Economic considerations may however favour Class 3 construction.

The Code, CP 110, devotes a large section to the specification and control of concrete mixes. For convenience, it recommends a series of grades for various categories of work; these are Grades 7, 10, 15, 20, 25, 30, 40, 50 and 60. The number associated with the grade is the specified characteristic strength of the concrete at 28 days. Throughout the Code, reference is made to the grades in the definition of requirements and the tabulation of the stresses to be used, and this procedure is followed in this and the later chapters.

Grades 30, 40, 50 and 60 are given in the Code as being appropriate for prestressed concrete work. In practice, concretes of Grades 40 and 50 are most likely to be used with a tendency for concrete of the higher strength to be employed with pretensioning since high early strength is required to effect a rapid turn round of equipment and plant. Concretes of lower strength may find some applications for post-tensioning but the higher losses of prestress due to creep which must be allowed for, as noted in Table 2.3, in concretes with a strength at transfer of less than 40 N/mm² discourage their use.

The compliance procedures for controlling the quality of the concrete in the factory or on the site require the concrete test cubes to be stored in water at 20° C until they are tested at 28 days. These results therefore provide no guidance to the strength of concrete at transfer and additional cubes must be stored with the units or structure to ensure that the strength needed at transfer is attained. Limitations on stresses at transfer are set out in Table 2.5 where it should be noted that, whilst compressive stresses are related to strength at transfer, the tensile stresses for structures of Classes 2 and 3 are somewhat illogically related to the characteristic strength, i.e. the Grade of concrete.

Table 2.5 *Allowable stresses in concrete at transfer*

Compression

     $0.5\, f_{ci}$ for triangular or near triangular distribution of prestress;
     $0.4\, f_{ci}$ for uniform or near uniform distribution of prestress;
         where, f_{ci} is the strength of the concrete at transfer.

Tension

Class 1 structures 1 N/mm^2

Class 2 and 3 structures

     Pretensioning $0.45\, (f_{cu})^{\frac{1}{2}} \text{ N/mm}^2$
     Post-tensioning $0.36\, (f_{cu})^{\frac{1}{2}} \text{ N/mm}^2$

     where, f_{cu} is the characteristic strength, i.e. the grade of concrete, in N/mm².

(For Class 3 structures higher stresses are allowed but the concrete section should then be assumed to be cracked in the calculations.)

The allowable stresses in the concrete in service are tabulated in Table 2.6. It will be seen that the compressive stresses at transfer are relatively higher in relation to the strength of the concrete at that stage than those adopted for design for service conditions. The main reasons for this are that the forces at transfer reduce with time as losses of prestress occur and at this stage the strength of the concrete is increasing rapidly. In service however, the loads do not usually reduce with time and further increases in the strength of the concrete are relatively small.

Table 2.6 *Allowable stresses in concrete in service*

Compression

$0.33f_{cu}$ in flexure which may be increased to $0.4f_{cu}$ in the region of support moments in statically indeterminate structures;

$0.25f_{cu}$ in direct compression;

$0.5f_{cu}$ in the prestressed concrete unit in composite construction at the interface of the *in situ* concrete provided that failure of the whole section would be in flexure in tension;

where, f_{cu} is the characteristic strength of the concrete.

Tension

Class 1 structures   None

Class 2 structures

Pretensioning	$0.45\,(f_{cu})^{\frac{1}{2}}$ N/mm^2
Post-tensioning	$0.36\,(f_{cu})^{\frac{1}{2}}$ N/mm^2

These stresses may be increased by up to 1.7 N/mm^2 provided that

(a) the tensile stress is less than three-quarters of that for cracking in tests on prototypes,

(b) the prestress in compression after all losses is not less than 10 N/mm^2,

(c) tendons or reinforcement are distributed in the tensile zone to restrain the development of cracking in the event of accidental overloading.

Class 3 structures

Grade of concrete	*Hypothetical allowable tensile stresses*/N/mm^2 *for members of* 0.4 m *in depth corresponding to the nominal crack widths given in* mm *for*			
	Pretensioned and grouted post-tensioned tendons		*Pretensioned tendons distributed in the tension zone close to the tension face*	
	0.1 mm	0.2 mm	0.1 mm	0.2 mm
30*	3.2	3.8	—	—
40	4.1	5.0	5.3	6.3
50 and over	4.8	5.8	6.3	7.3

* Post-tensioning only

Table 2.6—*continued*

For members of other depths the stresses given above should be multiplied by the following factors

Depth/metres	*Factor*
0·2 and less	1·1
0·2	1·0
0·6	0·9
0·8	0·8
1·0 and over	0·7

Where there is up to 1% of reinforcement in the section, the allowable tensile stresses obtained above may be increased linearly by up to 4 N/mm² for construction covered by the first two columns of the table and by up to 3 N/mm² for that covered by the second two columns. The value of this enhanced allowable tensile stress should not exceed $0.25f_{cu}$.

In situ concrete in composite construction (see Chapter 6)

Grade of in situ *concrete*	*Allowable tensile stress*/N/mm²
25	3·2
30	3·6
40	4·4
50	5·0

These stresses may be used in the *in situ* concrete at the contact surface with the precast concrete. They may be increased by up to 50% provided that the allowable tensile stress for the prestressed unit is reduced by the same amount of stress.

The allowable tensile stresses in the concrete in service given for structures of Classes 1 and 2 are realistic since the concrete is not cracked and is capable of sustaining them. Those given for Class 3 structures are described as hypothetical since they are generally in excess of the tensile strength of the concrete and it is therefore cracked. Nevertheless, in default of a better method of calculation which might well be more complex, limitations are imposed on the levels of equivalent tensile stress, which have been found by experiment to satisfy requirements for crack limitation, and it is assumed that the section is uncracked. One consequence is that if these assumptions are adopted in the calculation of deflections, then the deformation of Class 3 structures may be much underestimated. If the maximum tensile stresses allowed for Class 3 structures are developed, then it is possible for the stress across the section under the maximum design load for serviceability to vary from $0.33f_{cu}$ in compression to $0.25f_{cu}$ in tension. In these circumstances, it is also likely that deflection may be an important limitation in considering serviceability and that requirements for the ultimate limit state might be overriding.

2.11 DEFLECTION

The calculation of the deflection of prestressed concrete members to meet the requirements for the serviceability limit state presents fewer difficulties than for reinforced concrete construction since, apart from some Class 3 structures, the concrete is uncracked in normal service. Thus it is easier to set out formulae for a direct check on limit state requirements for deflection, but it is still only possible to give approximate estimates for actual structures. The main reasons for this, apart from the difficulty in being precise in estimating the deformation of concrete as briefly considered in section 2.7, are that:

(a) the simple conditions of support for beams and other members usually assumed in design seldom, if ever, exist in practice;
(b) the assumptions made for the magnitude and distribution of imposed load bear little direct relationship to reality;
(c) the loading history following transfer, when the concrete is still relatively immature and the members are being incorporated in the structure, cannot be forecast.

For a simply supported beam, which is uniform along its length and subjected to the prestress only, the final mid-span *upward* deflection is given by

$$\frac{P_e e l^2}{8 E_{ce} I} \tag{2.11}$$

where, P_e is the force in the tendons after all losses of prestress
e is the eccentricity of P_e, l is the span
I is the second moment of the area of the section, which may be taken for convenience as that for the concrete section
E_{ce} is the effective modulus of the concrete, which is given by

$$\frac{E_c}{1 + K_c E_c}$$

where E_c is the modulus of elasticity of concrete as given in Table 2.1 and K_c is the creep of concrete under unit stress as given in Table 2.3.

If the member is uniformly loaded so that the maximum moment is M, the final downward deflection may be estimated from:

$$\frac{5Ml^2}{48 E_{ce} I} - \frac{P_e e l^2}{8 E_{ce} I} = \left(\frac{5M}{48} - \frac{P_e e}{8} \right) \frac{l^2}{E_{ce} I} \tag{2.12}$$

Thus the deflection of a prestressed concrete member is the difference between the upward deflection due to the prestress and the downward deflection due to the load. The deflection at mid-span may be zero in certain circumstances, and this occurs theoretically for a rectangular section with an eccentricity of one-sixth of the depth when $M/M_{max} = 0.6$, where M_{max} is the moment corresponding to zero tensile stress in the concrete. For beams designed for bending tensile stresses under the design loads for serviceability, e.g. Class 2 structures, the moment for zero mid-span deflection would be a smaller ratio of the design moment, but not negligible and not necessarily far removed from the permanent moment in reality. In consequence the deflection of most beams in Classes 1 and 2 does not need to be checked in design. Where it is necessary to calculate deflections, particularly for continuous construction in prestressed concrete, the moment-area method has several advantages since it can be readily applied to the different combinations of prestress and loading that need to be considered and to beams with bent up tendons.

As already pointed out however, consideration of deflection may be necessary in the design of Class 3 structures. If the permanently imposed load is greater than one-quarter of the design imposed load, it should be assumed that the member is likely to be cracked in service and the limitations on span/depth ratio recommended for reinforced concrete beams should be applied. Otherwise the deflection should be estimated in the manner outlined above.

Since the deformation of most prestressed concrete construction is not accompanied by cracking and the tendons behave almost elastically over a wide range of stress, prestressed concrete members not only deform almost elastically but also show complete recovery from short term loading. Their performance under dynamic loading is appreciably better therefore than that of reinforced concrete members so long as the deformation is not excessive. If however the deformation leads to damage to the concrete or results in the tendons approaching their ultimate strength, performance is less satisfactory than for reinforced concrete, which has a much greater capacity for absorbing energy.

2.12 OTHER FEATURES AFFECTING SERVICEABILITY

Portland cement concrete provides an alkaline environment to the tendons which protects the steel against corrosion. With time, however, exposure of the concrete to the atmosphere leads to carbonation of the hydrated cement which progresses into the concrete from the surface and reduces its capability for preventing the steel from rusting. The thickness of cover of concrete given to the tendons in the recommendations in CP 110 is therefore related to the severity of the environmental conditions and the quality of the concrete, the quality being expressed in terms of its grade with some restrictions on water/cement ratio and cement content. The recommendations of CP 110 for cover to tendons in prestressed concrete also apply to any reinforcement incorporated in the section irrespective of whether it contributes to the loadbearing capacity of the member or not; these recommendations are given in Table 2.7.

Table 2.7 *Cover requirements for tendons and reinforcement in prestressed concrete*

Condition of exposure	*Minimum cement content**/kg/m³	*Maximum free water/cement ratio*†	*Nominal cover*/mm *for concrete of*		
			Grade 30	*Grade* 40	*Grade* 50 *and over*
mild	300	0·65	15	15	15
moderate	310	0·55	30	25	20
severe	370	0·45	40	30	25
very severe	—	—	—	60	50
subject to de-icing salt	310	0·55	50‡	40‡	25

* For maximum size of aggregate of 10 mm.
† Where water/cement ratio can be strictly controlled.
‡ For concrete with entrained air only.

Mild	completely protected against weather or aggressive conditions except for exposure to normal weather conditions briefly during construction
Moderate	sheltered from severe rain and against freezing whilst saturated with water—buried concrete and concrete under water
Severe	exposed to driving rain, alternate wetting and drying and freezing whilst wet—subject to heavy condensation or corrosive fumes
Very severe	exposed to sea water or moorland water with abrasion

Since the location of the steel in prestressed concrete is governed mainly by the need to create a particular distribution of stress in the section and not by the requirement to obtain the maximum possible lever arm as in reinforced concrete, the provision of adequate cover is less critical. This is particularly important since many structures require specific resistances to fire, which calls for greater amounts of cover to the tendons than are necessary for durability alone; this aspect is dealt with later in section 3.5 under ultimate limit states.

The spacing of tendons or their ducts within the concrete section is not the subject of detailed recommendations, but generally should be such that the concrete can be fully compacted around and between them without undue difficulty.

Since all tendons are of high tensile steel and tend to have a small cross-section, they are more susceptible to corrosion than the reinforcing bars used in ordinary reinforced concrete. Hence greater care is needed in their handling, storage and treatment to ensure that no pitting due to rusting occurs before stressing; when incorporated in the structure, they should be surrounded by dense concrete or grout; and in view of its potential for causing corrosion it is now generally recommended that calcium chloride should not be used in concrete or grout in direct contact with the tendons.

3

Analysis for Ultimate Limit States

3.1 BASIS FOR ANALYSIS

Design for the conditions at the ultimate limit state deals with the strength of the structure as a whole as well as the strength of its component parts. It is concerned with the need to provide acceptable levels of risk for whatever form failure may take, irrespective of whether the structural incidents that could cause failure are foreseen or not. Inevitably the treatment of the unexpected hazard is arbitrary but is an essential part of design particularly for buildings where the requirement is satisfied by the provision of vertical and horizontal steel ties as set out in CP 110. This aspect is not considered further here, and the method of analysis now given deals with the distribution of moments and forces due to dead, imposed, and wind loads and the resulting stresses in the materials.

The design loads for the ultimate limit state are obtained from the characteristic·loads as given by CP 3, Chapter V, multiplied by the values for γ_f in Table 1.1 for each of the combinations of loading that need to be considered. When dealing with serviceability limit states, it was necessary to examine all stages of manufacture and construction, but for the ultimate limit state it is only necessary to deal with the completed structure. The combinations of loading that must be allowed for are those which cause the most adverse conditions; for continuous beams these are assumed to correspond to alternate spans being loaded to $1{\cdot}4G_k + 1{\cdot}6Q_k$ in addition to the dead load G_k on all spans, or to adjacent spans loaded in this way, whichever results in the most adverse moments and shears at any section.

Although concrete is a brittle material, when acting in combination with steel reinforcement in reinforced concrete, it can be assumed to have limited plasticity. Prestressed concrete is less capable of deformation because the concrete is of higher strength which increases its capacity for elastic deformation without increasing its overall deformability, the tendons are of high tensile steel which is less ductile than mild steel, and the initial strain in the steel due to the prestress in effect reduces its ductility further. The design of statically indeterminate structures in prestressed concrete is based on elastic analysis for determining the distribution of moments and forces with some redistribution of moments but to a lesser degree than for reinforced concrete, as explained in Chapter 7.

3.2 STRENGTH IN FLEXURE

The Code, CP 110, offers two methods for calculating the ultimate flexural strength of prestressed concrete members. In one approach idealized forms for the stress–strain characteristics of the materials are assumed as shown for concrete in Figure 3.1, for reinforcement in Figure 3.2(a) and for tendons in

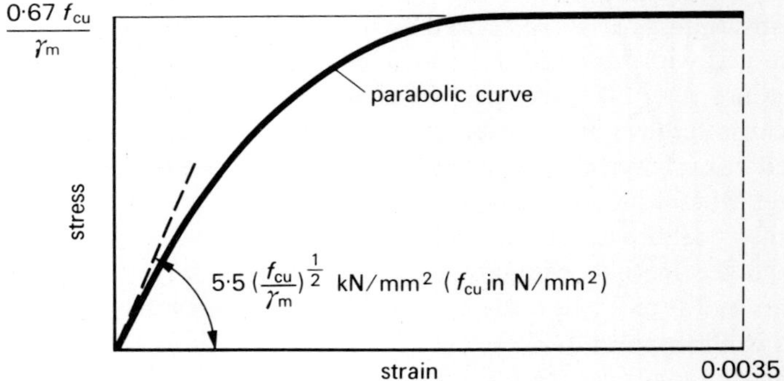

Figure 3.1 Stress–strain relationship for concrete

Figure 3.2(b). The strength is calculated by equating forces and moments whilst maintaining compatability of the strains in the materials. This method of calculation can also provide the relationship between moment and curvature for intermediate levels of loading should it be required for estimating deformation at loads in excess of those expected in normal service. The approach has been used in the preparation of the design charts which form Parts 2 and 3 of CP 110.

The other method of calculation, which is supported by extensive experimental study, is empirical and is the method now presented in some detail.

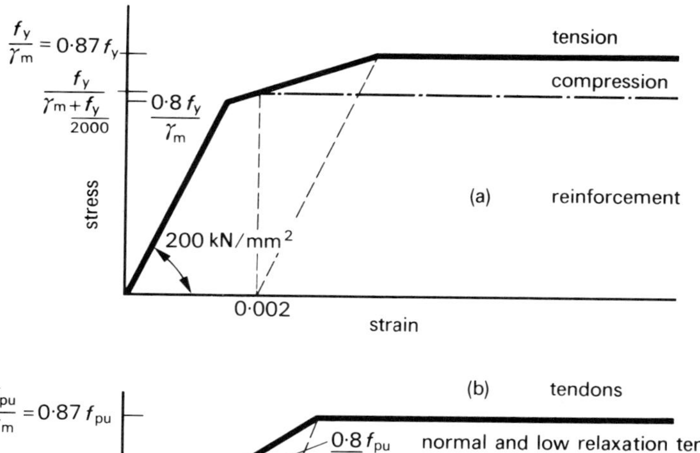

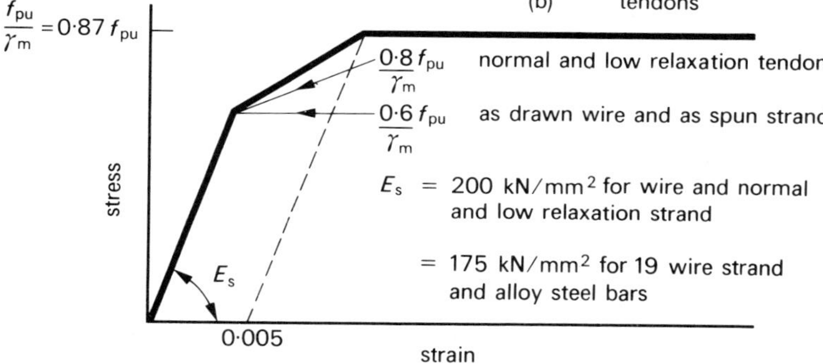

Figure 3.2 Stress–strain relationships for reinforcement and tendons

The following assumptions are made:

(a) plane sections before bending remain plane after bending;
(b) the stress in the concrete is uniformly distributed across the compression zone with a value of $0.6f_{cu}$ which is divided by γ_m for concrete to give a value of $0.4f_{cu}$ for use in the calculations;
(c) the ultimate compressive strain for concrete is 0.0035;
(d) the tensile strength of the concrete is ignored;
(e) the residual prestress in the steel is not less than 0.45 times its characteristic strength;
(f) the stress in the tendons at failure of the section is obtained from data in Figures 3.3 and 3.4 which are based on CP 110 and refer to pretensioning and to post-tensioning with grouted cables. (In special circumstances tendons may be left ungrouted and reference should then be made to CP 110.)

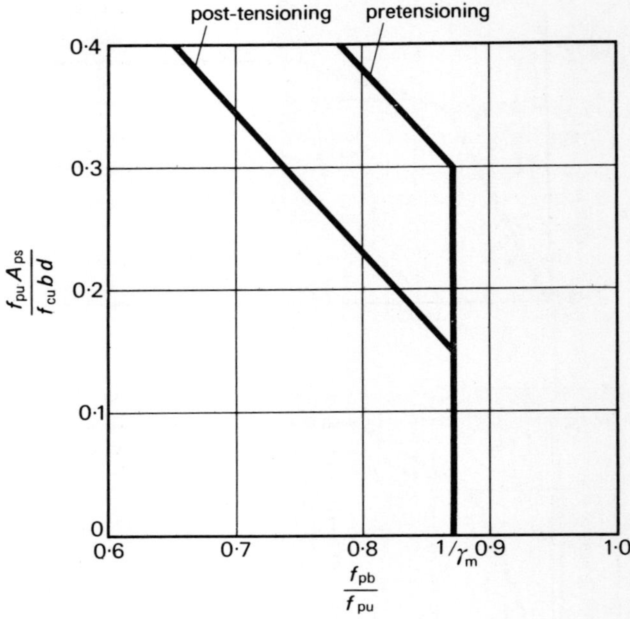

Figure 3.3 Stress in the tendons for beams at flexural failure

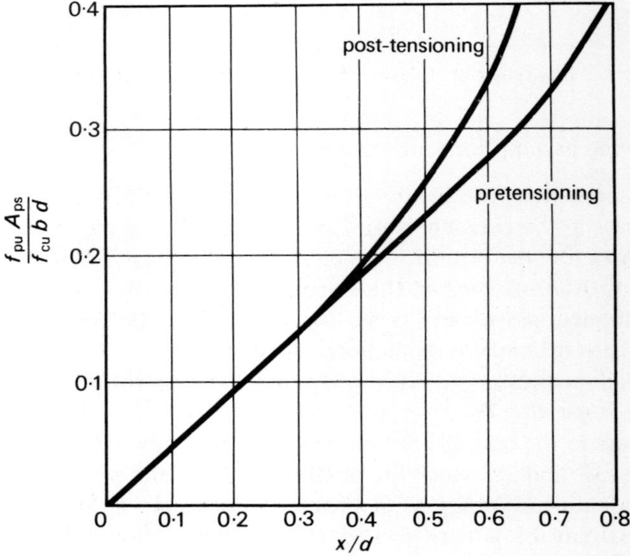

Figure 3.4 Depth of neutral axis for beams at flexural failure

The moment of resistance of the section at failure M_u (see Figure 3.5), is then given by

$$M_u = f_{pb}A_{ps}(d - 0.5x) \qquad (3.1)$$

where f_{pb} is the tensile stress in the tendons at failure and has the values given in Figure 3.3. It should be noted that the values for this stress have already been divided by γ_m for direct use in calculation.

A_{ps} is the area of the prestressing tendons

d is the effective depth

x is the depth of the neutral axis and has the values shown in Figure 3.4.

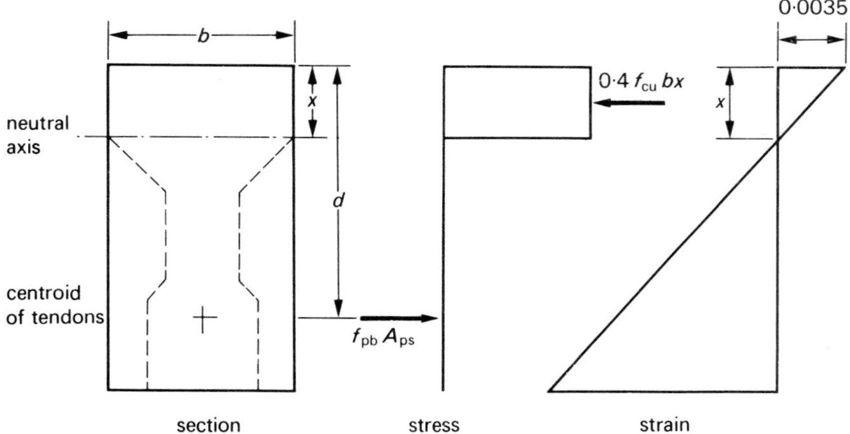

Figure 3.5 Distribution of stress in beams at flexural failure

This calculation is applicable only to beams with a rectangular cross-section above the neutral axis at failure. For other shapes of section, it should be assumed that the average stress in the compression zone is $0.4f_{cu}$ as before. Then the position of the neutral axis is determined by trial; a value for x/d must be selected in such a way that the total compression in the concrete is equal to the total tension in the tendons derived for that value of x/d from Figure 3.6. The ultimate strength is then the product of the total compression or tension force and the lever arm between the centres of compression and tension.

If there are tendons in the compression zone, their contribution to the initial moment at failure may be estimated from the distribution of stress in the section defined by the strain in the concrete at failure of 0.0035, as shown in Figure 3.5, the initial stress conditions, and the stress-strain relationship

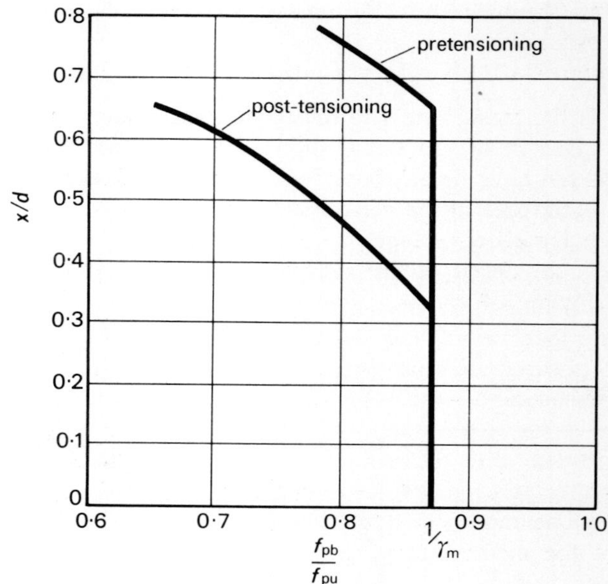

Figure 3.6 Stress in the tendons for different depths of the neutral axis at failure

for the tendons as given in Figure 3.2(b). The stress in any reinforcement can be determined in the same way using the stress–strain relationship in Figure 3.2(a).

Calculations to check the ultimate limit state requirements will usually be made on sections which have already been dimensioned to satisfy the requirements of the serviceability limit states. As a result, the value of the ratio $f_{pu}A_{ps}/(f_{u}bd)$ will already have been established and hence the values of f_{pb} and x can be determined from Figures 3.3 and 3.4 for substitution in (3.1) above. The assumptions used in the American Code ACI 318–71 are given in sections 10.3 and 18.7.

The information given so far for the calculation of the flexural strength of beams with post-tensioned tendons relates to grouted tendons. In most applications of prestressing, the tendons should be grouted in order to avoid dependence on the anchorage for the integrity of the member, to protect the tendon from corrosion and to develop a higher flexural strength through bond between the steel and the concrete. Also if tendons are not grouted, the members may become unstable if, as a result of accidental damage or in the course of demolition, the energy in the tendons is suddenly released. If it is necessary to calculate the flexural strength of beams with unbonded tendons, for example, if the adequacy of grouting is in doubt, the method given in CP 110 may be used or alternatively it may be assumed that the stress in the tendons at failure is 103 kN/mm² above the effective prestress as recommended in the ACI code.

3.3 STRENGTH IN SHEAR

The recommendations of CP 110 require that beams and slabs should be designed for shear resistance at the ultimate limit state only, with the exception of one aspect of the design of composite beams for shear, which will be included in this chapter for convenience.

At ultimate load beams and slabs may be subjected to the most severe conditions of shear at sections which are not cracked in flexure; on the other hand, the presence of a point load may cause the most severe conditions of moment and shear at the same section. Hence CP 110 gives methods of calculation to deal with the shear resistance of sections with and without flexural cracking. In each case an estimate is made of the shear strength of the concrete alone and of the contribution it may make to the total required resistance of the member; the remaining shear is then carried by shear reinforcement.

For sections of beams, including composite beams, and slabs which are not cracked in flexure at ultimate load, the method of calculation is based on restricting the value of the maximum principal tensile stress at the geometric axis of the section to $0 \cdot 24(f_{cu})^{\frac{1}{2}}$ in N/mm² units. The ultimate shear resistance V_{co} of the concrete section, is then derived as given below.

The principal tensile stress f_t at the centroid of the section (taken as positive), is given by

$$f_t = -\left(\frac{f_{cp}+f_v}{2}\right)+\left[\left(\frac{f_{cp}-f_v}{2}\right)^2+v^2\right]^{\frac{1}{2}} \tag{3.2}$$

where f_{cp} = the longitudinal prestress at the centroid of the section (taken as positive),

f_v = the vertical prestress (taken as positive),

v = $1 \cdot 5 V_{co}/bh$ for a rectangular section with a breadth b and depth h, i.e. shear stress at the centroid of the section.

Substituting for v leads to:

$$V_{co} = 0 \cdot 67bh(f_t{}^2+f_tf_{cp}+f_tf_v+f_{cp}f_v)^{\frac{1}{2}} \tag{3.3}$$

Since a γ_m value of $1/0 \cdot 8$ has been applied in defining f_t as $0 \cdot 24(f_{cv})^{\frac{1}{2}}$, it seems logical to apply the same factor to the prestress components f_{cp} and f_{cv}.

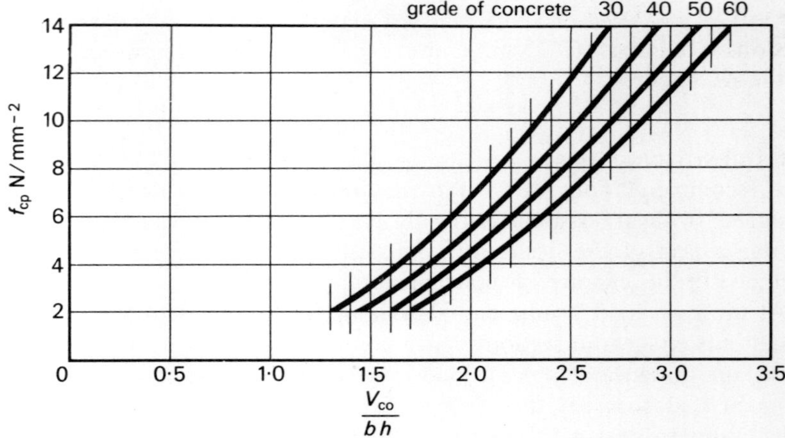

Figure 3.7 Ultimate shear resistance of the uncracked concrete section

Hence

$$V_{co} = 0 \cdot 67bh(f_t^2 + 0 \cdot 8f_t f_{cp} + 0 \cdot 8f_t f_v + 0 \cdot 64f_{cp}f_v)^{\frac{1}{2}} \tag{3.4}$$

If the vertical prestress is zero then

$$V_{co} = 0 \cdot 67bh(f_t^2 + 0 \cdot 8f_t f_{cp})^{\frac{1}{2}} \tag{3.5}$$

as in CP 110. b is also taken as the breadth of the web for I, T and L sections since the approximation is small.

Values for V_{co} for different values of f_{cp} obtained from (3.5) are plotted in Figure 3.7. In beams with inclined tendons, the vertical component of the prestress in the tendons imposes shear on the section which normally counteracts the shear due to dead and imposed loads and in these circumstances may be deducted from it in the design calculations. If a vertical prestress is introduced by vertical tendons, then expression (3.4) should be used.

The method of design for sections cracked in flexure takes account of the possibility that flexural cracks may ultimately develop into inclined shear cracks and provides for a transition in determining the shear strength of the concrete section from the reinforced concrete member with no prestress at one limit to the Class 1 prestressed concrete member, which receives the maximum assistance in shear from the prestress, at the other. In this way, the shear resistance of Class 2 beams and particularly Class 3 beams, which may have little prestress, is embraced. The ultimate shear strength V_{cr} of the concrete section cracked in flexure is given by

$$V_{cr} = \left(1 - 0\cdot55\,\frac{f_{pe}}{f_{pu}}\right) v_c bd + M_0 \frac{V}{M} \tag{3.6}$$

where f_{pe} is the effective prestress in the tendons after all losses and should not be taken as greater than $0\cdot6f_{pu}$, f_{pu} being the characteristic strength of the tendon;

v_c is the ultimate shear stress in the concrete as for reinforced concrete which is plotted in Figure 3.8;

b is the breadth of the section as in the previous paragraph;

d is the effective depth;

M_0 is the moment necessary to produce zero stress in the concrete at depth d assuming that the prestress at this depth is $0\cdot8$ of that obtained by calculation;

V and M are the shear and moment at ultimate load at the section considered;

V_{cr} may be given a value of not less than $0\cdot1bd(f_{cu})^{\frac{1}{2}}$, and the value calculated from (3.3) may be assumed constant for a distance of $d/2$ in the direction of increasing moment from the section.

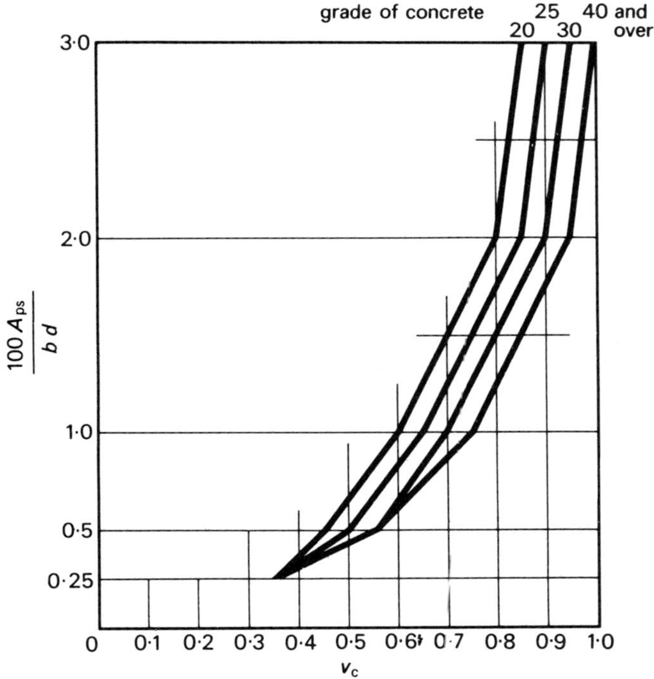

Figure 3.8 Ultimate shear stress v_c for the cracked concrete section

Having determined V_{cr} for a section cracked in flexure, a check should be made for resistance of the concrete in the absence of cracking by determining V_{co}. The lower of these two values should be used in the design. If the beam has inclined tendons where the section is cracked, the vertical component should not be deducted from the shear due to the dead and imposed load, since experimental evidence has shown that using inclined tendons leads to more extensive cracking and hence reduces the shear resistance of the concrete section.

No shear reinforcement is needed in floor slabs, footings, pile caps or members of minor importance when the shear V is less than V_c (either V_{co} or V_{cr} as appropriate). For other members, shear reinforcement is not required when V is less than half V_c, and for these members when V is between a half V_c and V_c nominal reinforcement is necessary. The minimum amount is given by:

$$\frac{A_{sv}}{s_v} = \frac{0\cdot4b}{0\cdot87f_{yv}} \tag{3.7}$$

where f_{yv} is the characteristic strength of the reinforcement but not greater than 425 N/mm^2;

A_{sv} is the cross-sectional area of the two legs of a link;

s_v is the spacing of the links.

When V exceeds V_c, then shear reinforcement is required and should not be less than

$$\frac{A_{sv}}{s_v} = \frac{V - V_c}{0\cdot87f_{yv}d_t} \tag{3.8}$$

where d_t is the depth from the extreme compression face to the bars or tendons around which the links must pass close to the tensile face.

This relationship is similar to that used for shear in reinforced concrete and is derived by assuming that the excess shear $(V - V_c)$ must be resisted by the stirrups acting as vertical ties in a truss with the concrete providing the inclined compression members.

The links at a cross-section should enclose between them all the tendons and any additional reinforcement at the section. The spacing of the links should not be greater than $0\cdot75d_t$ or four times the web thickness along the member or $0\cdot75d_t$ laterally for the individual legs of the links. If V is greater than $1\cdot8V_c$, the longitudinal spacing should be reduced to $0\cdot5d_t$. The maximum shear force at ultimate load V should not exceed $0\cdot75(f_c)^{\frac{1}{2}}bd$ in N and mm units.

The principles of the American Code method are similar but there are differences in the detailed formulae which are given in ACI 318–71, Chapter 11.

The design of composite beams for shear follows that for ordinary prestressed concrete beams in the preceding paragraphs, except that shear at the interface between the precast and the *in situ* concrete is considered at the limit state of serviceability and therefore under the characteristic loads with γ_f equal to 1·0 for dead and imposed loads acting together. The horizontal shear stress at the interface is calculated from:

$$v_h = \frac{V_d S_c}{I b_e} \tag{3.9}$$

where v_h is the horizontal shear stress at the interface;

V_d is the total vertical shear at the section under the design load for the serviceability limit state;

S_c is the first moment of area of the concrete section to one side of the interface about the centroid of the composite section;

I is the second moment of area of the composite section;

b_e is the width of the interface.

The allowable horizontal shear stresses for three types of surface between the precast and the *in situ* concrete taken from CP 110 are given in Figure 3.9 for beams. For composite sections reinforced for resisting vertical shear, this reinforcement should extend into the *in situ* concrete and may then be used to resist horizontal shear. For composite floor slabs without links the hori-

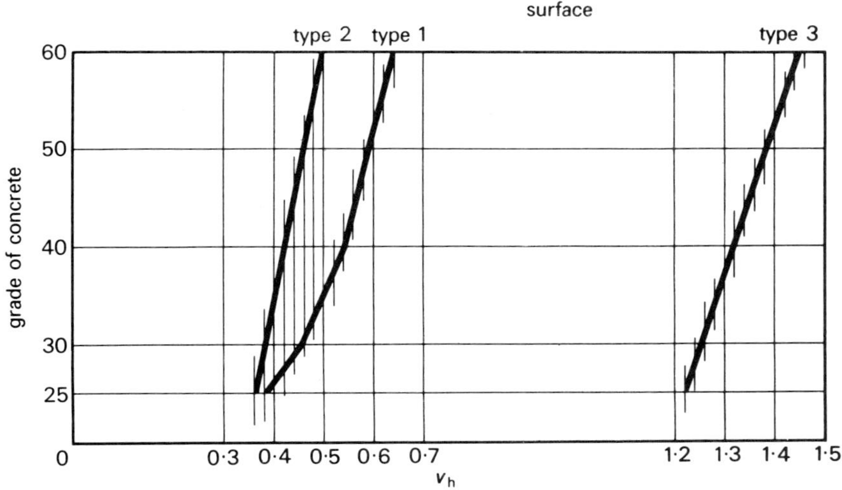

Figure 3.9 Allowable shear stresses for shear connection in composite beams

zontal shear stresses for Type 1 surfaces should not exceed $0 \cdot 7 \text{ N/mm}^2$ or $1 \cdot 2$ times the corresponding value for beams in Figure 3.9 and, where there is no special surface treatment, $0 \cdot 8$ times the stresses for Type 2 surfaces. The application of this method is discussed in Chapter 6.

3.4 END ANCHORAGES FOR POST-TENSIONED TENDONS

At the end of each post-tensioned tendon, a concentrated load is applied to the concrete through the anchorage device which usually consists of an embedded cone or cylinder. In the immediate vicinity there exists a complex system of stresses in the two directions transverse to the direction of the prestressing force. These stresses may be sufficient to cause cracking of the concrete, and in the absence of reinforcement a catastrophic failure may ensue. In design, the usual procedure is to assess the total force ('the bursting force'), in each transverse direction and to provide reinforcement of appropriate strength.

The Code of Practice gives a table of recommended design values of the bursting force resulting from an axial force applied by a tendon to a square concrete end block (CP 110, Table 43). The bursting force is proportional to

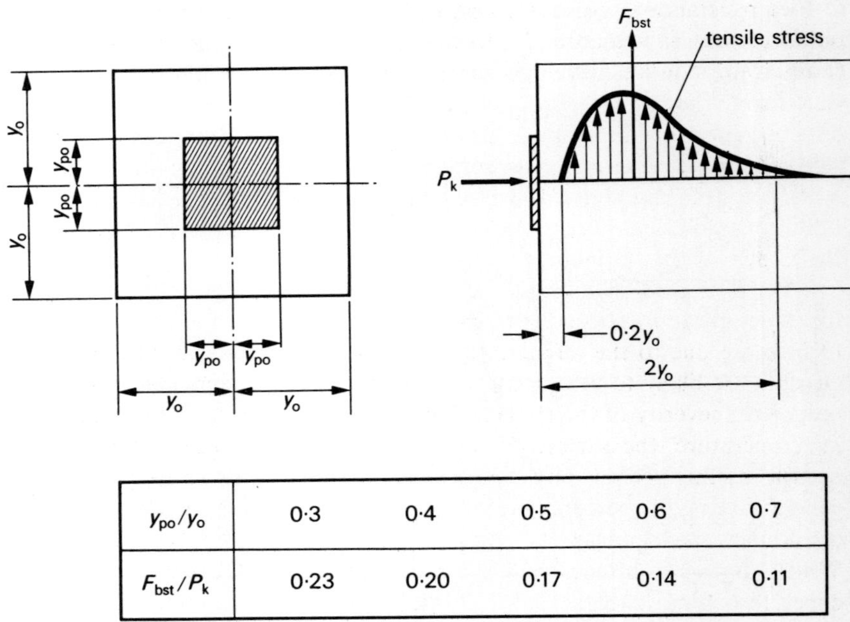

y_{po}/y_o	0·3	0·4	0·5	0·6	0·7
F_{bst}/P_k	0·23	0·20	0·17	0·14	0·11

Figure 3.10 Design bursting tensile forces in end blocks (CP 110)

the force applied P_k and also depends on the size of the anchorage in relation to the end of the block (Figure 3.10), as represented by the ratio y_{po}/y_o.

P_k is the tendon jacking load or the load in the tendon at the ultimate limit state, whichever is the greater. The jacking load will normally be greater, but if the tendon is unbonded it may sometimes be exceeded by the load at the ultimate limit state. The reinforcement is assumed to act at its design strength of $0.87 f_y$, but the stress is limited to a value corresponding to a strain of 0.001 when the concrete cover is less than 50 mm.

Multiple anchorages are treated by dividing the end block into a series of symmetrically loaded prisms, each of which is considered as described above.

3.5 FIRE RESISTANCE

All construction governed by building regulations must satisfy requirements for fire resistance which cover the retention of strength in fire, resistance to the penetration of flame and resistance to the transmission of heat. So far as the structure of a building is concerned, the method for judging adequacy is by full-scale fire test, but since testing facilities are limited in size the data relate to the performance of individual members in isolation from the remainder of the structure. Hence an important effect is omitted from the tests and the application to real structures is limited.

Fire resistance is considered here in relation to the ultimate limit states since, although the loading for the test conditions is that for the serviceability limit state, failure develops when the strengths of the materials have deteriorated under the effects of heat to such an extent that collapse occurs. Fire resistance is defined in terms of the time for which the structural member remains intact supporting its service load when subjected to heating to a standard time–temperature curve. The required duration is given in the building regulations according to the size of the building or compartment and the nature of the occupancy.

When a prestressed concrete beam is heated from below there is a tendency for its length to increase due to the rise in temperature and for the deflection to increase due to the temperature gradient. In continuous structures, these changes are likely to produce a redistribution of moments and forces and to reduce the severity of the effects of the fire on the member. With the increase in temperature, the surface of the concrete dries and fine surface cracking occurs; as this process continues the cracking becomes more extensive and general drying of the concrete leads to shrinkage which to some degree may counteract the changes in length and deflection caused by temperature. Whilst drying from the heated surfaces continues, the temperature in the concrete will not exceed $100°$ C. When however the concrete has dried near the heated surface, its temperature rises rapidly and eventually the temperature in the tendons and any reinforcement will also rise rapidly. Failure will occur when the steel has been heated to such an extent that its strength has reduced below that necessary to support the load. The effect of temperature on the strength of the concrete in the compresson zone is much less significant. Design for fire resistance is therefore primarily concerned with providing enough protection to the steel tendons.

When prestressing tendons of hard drawn steel wire are heated to about 400° C they lose about half their strength. By comparison the yield stress of most reinforcing steels drops to about half their strength at room temperature when heated to about 600° C, and therefore the protection needed for the steel in prestressed concrete is appreciably greater than for reinforced concrete. Since the overall load factor corresponding to a combination of γ_m and γ_f is between 1·7 and 2·0, a prestressed beam is likely to fail when the strength of the steel reduced to about a half; failure is therefore to be expected when the tendons reach a temperature of about 400° C. Since Class 3 structures may be subjected to a greater range of stress under service loads than other structures, they may tend to failure when the tendons reach a slightly lower temperature.

Protection to the steel to delay the rise in temperature is obtained by choosing an appropriate thickness of concrete cover which may be augmented by the addition of insulating material such as sprayed asbestos, vermiculite concrete or plaster. The thickness of concrete cover required is approximately 40 mm for 1 hour, 65 mm for 2 hours and 100 mm for 4 hours of fire resistance when natural aggregates are used. When siliceous aggregates are contained in the concrete there is a tendency for spalling to occur at high temperatures which is much less likely with calcareous or lightweight aggregates. With siliceous aggregates it has been found necessary to reinforce the cover with a light steel mesh when it is 45 mm or more. Concretes made with lightweight aggregates also have the advantage of having better insulating properties than those with natural aggregates and hence the concrete cover needed can be reduced by 20%.

Fairly extensive details of the requirements for fire resistance for a variety of forms of construction in prestressed concrete are given in CP 110 to which reference should be made. It should however be appreciated that the fire resistance requirements for building structures can have an important influence on their design and construction.

3.6 SECONDARY REINFORCEMENT

Reinforcement is necessary in prestressed concrete to provide shear resistance particularly in beams with thin webs, to permit the adoption of higher tensile stresses in the concrete of Class 3 structures, to counteract bursting tensile stresses in the anchorage region for post-tensioned tendons and to reinforce the concrete cover in meeting some requirements for fire resistance. Reinforcement may also be required in the form of links around pretensioned tendons along the transmission length to prevent bursting particularly if strand is used and steel ties may be needed to provide overall stability to the structure in the event of accidental damage. It may be desirable in beams prestressed by post-tensioning, when these are large of special shapes, to restrict cracking after casting caused by restraint of the formwork or by cooling of the concrete. Often this reinforcement can be used to serve a dual purpose, such as contributing to flexural strength whilst at the same time providing the longitudinal carrier bars for links necessary as shear reinforcement.

It should be recognized that the enhancement in strength gained by prestressing concrete is usually unidirectional and therefore a prestressed concrete member may be particularly vulnerable to loading or accidental damage for which it was not designed. There is much to be said therefore for incorporating a nominal amount of secondary reinforcement in any major prestressed concrete member.

C

4

Beams with Pretensioned Tendons

4.1 PRESTRESSING TECHNIQUE

In the pretensioning process, illustrated diagrammatically in Figure 4.1, the tendons, which normally consist of single or stranded cold-drawn wires, are tensioned and temporarily secured to anchorages. The anchorages are usually separate from the member, Figure 4.1(a), although occasionally the tendons may be anchored to the mould if the latter is sufficiently stiff. The concrete is then placed, Figure 4.1(b), and when it has gained the necessary strength and is adequately bonded to the tendons, the latter are slowly released from the anchorages. The reaction of the tendons is thus transferred to the concrete member, Figure 4.1(c), the actual transmission of the force being effected by bond stress over a relatively short length at each end of the tendons, known as the transmission length.

Pretensioning is particularly appropriate for the manufacture of precast prestressed elements. It is ideally a factory operation, since very strong anchorages are required to withstand the total force in the tendons, but a single pair of anchorages may be used to produce a line of similar members. Moreover, factory conditions are most conducive to the good control and supervision necessary to obtain the specified strength of concrete in a short time (usually not less than one day), and thereby maintain a high rate of production.

There are, however, certain constraints imposed by pretensioning. As in all precast work, the size of element is limited by the capacity of handling equipment, both at the factory and at the site, and by transport considerations. The transmission of the prestress by bond restricts the diameter of the tendons and precludes the use of the very large tendons possible in post-tensioning.

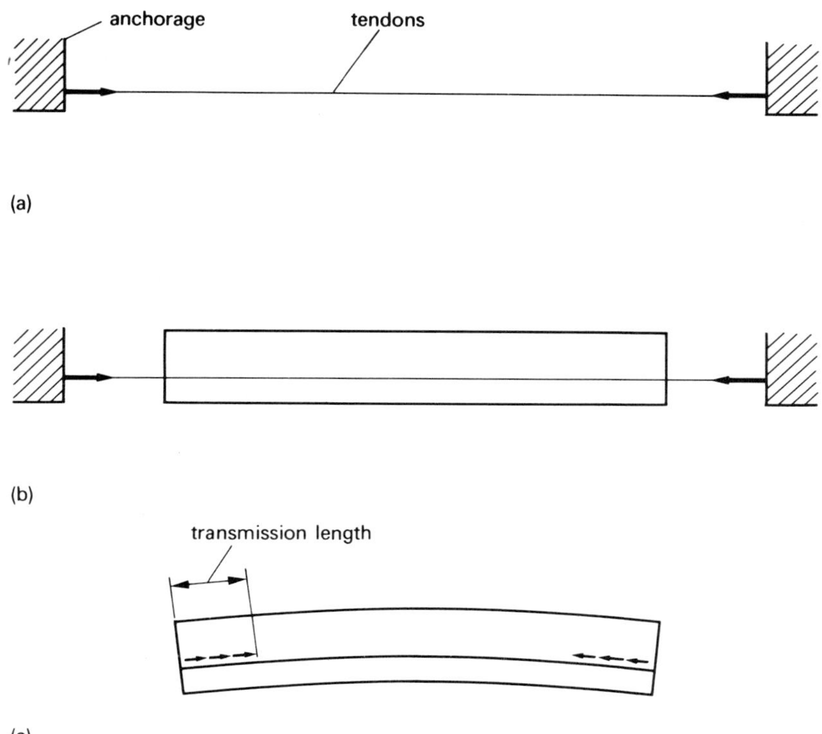

Figure 4.1 Pre-tensioning process: (a) tensioning; (b) casting; (c) prestressing concrete (transfer of prestress)

It is also less easy to vary the eccentricity of the prestressing force with respect to the axis of the beam, although this requirement may, to some extent, be met by deflecting the tendons on the pretensioning bed by means of large vertical forces as in Figure 4.2. As an alternative the prestressing force may be reduced by debonding part of the length of some tendons by inserting them in sleeves. The designer must bear in mind that the full prestress is not available at a distance less than the transmission length from the end of a member, and also that the losses of prestress due to elastic deformation of the concrete and to creep and shrinkage are somewhat larger than in post-tensioned tendons.

For the above reasons pretensioning is generally economical for prestressed elements of small to medium size, for example beams, of length up to about 20 m, while large structures tend to be prestressed *in situ* by post-tensioned tendons.

Figure 4.2 Pre-tensioning bed showing deflected tendons. (Photo by courtesy of Leonard Fairclough Ltd)

4.2 DIMENSIONING OF CONCRETE SECTION FOR SERVICEABILITY

The dimensions of the concrete section, the prestressed tendons and any additional non-prestressed reinforcement must be such that the ultimate resistance and the resistance at the serviceability limit are at least equal to the characteristic load multiplied by the specified factors. Particular dimensions tend to be controlled by one or other of the two limit states and it is possible to obtain a rapid solution by considering the two concurrently, refs. (4.1, 4.2). A more common alternative is to dimension the member for one limit state requirement, and then to check and if necessary amend the design for the other. In prestressed concrete design, as explained in Chapter 2, it is usually preferred to dimension in the first instance for the serviceability requirements, and to check for the ultimate limit state, this procedure will here be adopted.

The overall dimensions of the section will sometimes be determined by considerations other than strength and serviceability, but the minimum possible depth will obviously be controlled by the limit state requirements.

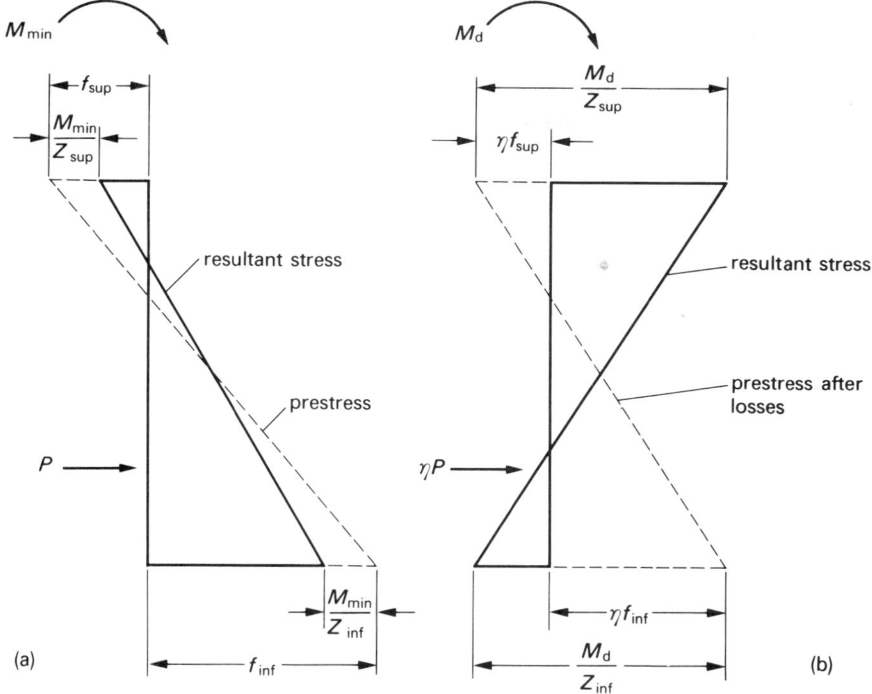

Figure 4.3 Critical combinations of prestress and bending stress due to load: (a) maximum prestress + minimum load; (b) minimum prestress + maximum (design) load

Deflection is less likely to be a governing factor than in reinforced concrete, and serviceability centres mainly on the relationship of the design load to the flexural cracking load, which depends on the stress in the concrete.

There are two combinations of prestress and load which are critical with respect to the stresses in the concrete at any section of a prestressed beam. When the maximum prestress, which exists when the concrete is first prestressed, is combined with the bending stress due to the minimum load as in Figure 4.3(a), the resultant compressive stress at the bottom of the beam is a maximum, as also is the tensile stress at the top. Secondly, when the minimum prestress, after all losses, is combined with the bending stress due to the maximum load, the resultant tensile stress at the bottom and the resultant compressive stress at the top attain their maximum values [Figure 4.3(b)]. There is an exception where the prestress at the top of the beam is compressive; under this condition the greatest resultant compressive stress is produced by a combination of the maximum prestress and the maximum load.

Here the maximum load is the design load for the serviceability limit state. The minimum load is that which produces the least positive moment at the section in question, and is normally the self-weight of the prestressed beam. Although it is usual to consider the self-weight acting on the span for which the beam is designed, it should be appreciated that more severe conditions may occur for a short period when the beam is lifted at points other than the bearings.

It will be seen from Figure 4.3 that the following four stress conditions have to be fulfilled in the concrete.

$$f_{inf} - \frac{M_{min}}{Z_{inf}} \leqslant f_{cp\,adm} \tag{4.1}$$

$$f_{sup} - \frac{M_{min}}{Z_{sup}} \leqslant f_{tp\,adm} \tag{4.2}$$

$$\frac{M_d}{Z_{inf}} - \eta f_{inf} \leqslant f_{t\,adm} \tag{4.3}$$

$$\frac{M_d}{Z_{sup}} - \eta f_{sup} \leqslant f_{c\,adm} \tag{4.4}$$

or when f_{sup} is a compressive stress:

$$\frac{M_d}{Z_{sup}} - f_{sup} \leqslant f_{c\,adm} \tag{4.4a}$$

where

M_d = design moment for serviceability limit state

M_{min} = minimum moment

$f_{cp\,adm}$ = allowable compressive stress in concrete at the time of prestressing

$f_{t\,adm}$ = allowable tensile stress in concrete at age when design moment is first applied

$f_{c\,adm}$ = allowable compressive stress in the concrete at age when design moment is first applied

$f_{tp\,adm}$ = allowable tensile stress in concrete at the time of prestressing.

The values of the allowable stresses have been considered in Chapter 2. They are specified in CP 110, 4.3.3 or in the American Code ACI 318–71, 18.4.

$$\eta = \text{loss ratio} = \frac{\text{final value of prestress}}{\text{prestress first applied to concrete}}$$

(4.1) and (4.3) refer to the bottom of the section, (4.2) and (4.4) to the top. Eliminating the prestress f_{inf} from (4.1) and (4.3) yields the following formula for the minimum value of the section modulus required for the bottom of the section, given the loading and allowable stresses in the concrete:

$$Z_{\text{inf}} \geqslant \frac{M_{\text{d}} - \eta M_{\text{min}}}{\eta f_{\text{cp adm}} + f_{\text{t adm}}} \tag{4.5}$$

A corresponding formula for the minimum section modulus required for the top of the section may be derived from equations (4.2) and (4.4):

$$Z_{\text{sup}} \geqslant \frac{M_{\text{d}} - \eta M_{\text{min}}}{f_{\text{c adm}} + \eta f_{\text{tp adm}}} \tag{4.6}$$

or when the prestress at the top is compressive:

$$Z_{\text{sup}} \geqslant \frac{M_{\text{d}} - M_{\text{min}}}{f_{\text{c adm}} + f_{\text{tp adm}}} \tag{4.6a}$$

Formula (4.5) is particularly useful for the preliminary dimensioning of beam sections, at which stage the load is known approximately, the allowable stresses have been specified, and the loss ratio may be estimated. A suitable section, having a section modulus greater than the value of Z_{inf} indicated by the formula is selected from the available standard beam sections, or purpose-designed if necessary.

Although equation (4.6), [or (4.6a)] must also be satisfied, it is not the only criterion of the value of Z_{sup}. This is because the size of the top flange required to satisfy the ultimate limit state will usually result in a larger section modulus than that required to satisfy the stress conditions for the serviceability limit state. In addition, when the minimum load is relatively large, an adjustment of the prestressing force may be required, as explained later. This also requires an increased value of Z_{sup} which, to avoid later difficulties, should be made substantially greater than the value given by formula (4.6), [or (4.6a)].

4.3 STANDARD BEAM SECTIONS

It is usual to neglect the effect of the steel tendons and reinforcement on the dimensional properties of the section at this stage of design. The minimum dimensions of a rectangular beam cross-section having a required section modulus are therefore readily determined from the relationship:

$$Z_{\text{inf}} = Z_{\text{sup}} = \frac{bh^2}{6}$$

The rectangular cross-section, however, is rather inefficient in that the area, and hence the self-weight, is large in relation to the second moment of area, while the required prestressing force, which determines the amount of high strength steel in the tendons, is greater than that needed for a beam of I or T section designed for the same conditions. Sections of the latter type are

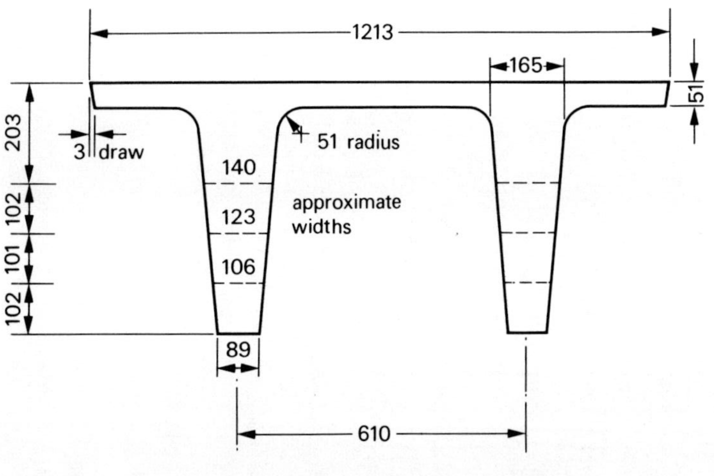

Figure 4.4 Standard double-T sections

Table 4.1 *Dimensional properties of double-T sections*

Section No.	Depth h/mm	Area A/mm²	Height of centroid y_{inf}/mm	Second moment of area I/mm⁴ × 10⁶	Section moduli		Self-weight kN/m²
					Z_{sup}/mm³ × 10⁶	Z_{inf}/mm³ × 10⁶	
DT 1	203	108 000	135	364	5·34	2·70	2·1
DT 2	305	134 000	200	1115	10·66	5·57	2·6
DT 3	406	157 000	264	2365	16·75	8·95	3·0
DT 4	508	177 000	332	4120	23·40	12·42	3·4

therefore preferable, except for unimportant members, and tables are available, refs. (4.1, 4.2) to facilitate dimensioning.

Several sizes of section are tending to become standard for certain types of construction, notably the Concrete Society double-T beams for floor construction, and the Cement and Concrete Association I and inverted T beams for composite bridge decks. The design of the former, the dimensions and section properties of which are given in Figure 4.4 and Table 4.1, is illustrated in the following example, while the design of composite beams will be discussed in Chapter 6.

Example

Selection of double-T section for floor construction

Precast concrete floor units are to be designed for a span of 9·5 m, to support a load of 1·4 kN/m² due to the floor finish together with an imposed load of 3·0 kN/m².

A span/depth ratio of 25 gives a rough indication of the depth:

$$h \simeq \frac{9500}{25} = 380 \text{ mm}$$

The loading may, therefore, be calculated assuming the use of the double-T unit DT3 (Table 4.1), the depth of which is 406 mm:

self-weight of unit DT3	3·0 kN/m²
floor finish	1·4 kN/m²
total dead load	4·4 kN/m²
imposed load	3·0 kN/m²
design load (serviceability limit state)	7·4 kN/m²

Moments on each unit, at midspan:

$$\text{moment due to self-weight} = 3 \cdot 0 \times 1 \cdot 213 \times 9 \cdot 5^2 \times 0 \cdot 125 = 41 \cdot 1 \text{ kNm}$$

$$\text{moment due to design load} = 7 \cdot 4 \times 1 \cdot 213 \times 9 \cdot 5^2 \times 0 \cdot 125 = 101 \cdot 2 \text{ kNm}$$

Formula (4.5) is now used to check the section modulus. Concrete of Grade 50 will be specified and will be assumed to have a strength of 35 N/mm² at the time of prestressing (transfer) and 50 N/mm² when the design load is first applied. According to CP 110 and designing the units as Class 2 members, the allowable stresses are therefore as follows (see also Tables 2.5 and 2.6):

compressive stress at transfer (Table 36)
$$f_{\text{cp adm}} = 0 \cdot 5 \times 35 = 17 \cdot 5 \text{ N/mm}^2$$

tensile stress at transfer (Table 33 extrapolating)
$$f_{tp\ adm} = 2\cdot7\ \text{N/mm}^2$$

compressive stress at design load (Table 32)
$$f_{c\ adm} = 0\cdot33 \times 50 = 16\cdot7\ \text{N/mm}^2$$

tensile stress at design load (Table 33)
$$f_{t\ adm} = 3\cdot2\,\text{N/mm}^2$$

The estimated value of the loss ratio η is 0·75. If desired a more accurate estimate may be made as explained in Chapter 2.

The value of the minimum moment M_{min} to be used in formula (4.5) will depend on whether the calculation applies only to the critical section at one point on the beam (e.g. midspan) and it is possible to vary the section and/or the prestress at other points by the use, for example, of deflected or debonded tendons. If this is so M_{min} will normally be the moment at the given point due to the self-weight of the beam. If, however, the beam is of uniform section with the tendons parallel to the axis, so that the prestress is to be the same at all points, it is necessary to consider the maximum and minimum moment for all points on the beam, and for a simply supported beam M_{min} will be zero at the supports. The two cases will be considered separately.

(a) Tendons parallel to axis.

$$\text{From formula (4.5)} \quad Z_{inf} \geqslant \frac{(101\cdot2 - 0\cdot75 \times 0) \times 10^6}{0\cdot75 \times 17\cdot5 + 3\cdot2}$$

$$= 6\cdot25 \times 10^6\ \text{mm}^3$$

The standard double-T section DT3, for which $Z_{inf} = 8\cdot95 \times 10^6\ \text{mm}^3$, is sufficient (Table 4.1).

(b) Deflected or debonded tendons.

$$\text{From formula (4.5)} \quad Z_{inf} \geqslant \frac{(101\cdot2 - 0\cdot75 \times 41\cdot1) \times 10^6}{0\cdot75 \times 17\cdot5 + 3\cdot2}$$

$$= 4\cdot35 \times 10^6\ \text{mm}^3$$

Here the smaller section DT2 ($Z_{inf} = 5\cdot57 \times 10^6\ \text{mm}^3$) may be used.

In both cases the section modulus Z_{sup} is well in excess of the minimum value indicated by equation (4.6).

4.4 DESIGN OF TENDONS AT CRITICAL SECTION FOR SERVICEABILITY

When the concrete section has been selected the values of Z_{inf} and Z_{sup} are known, and the maximum and minimum values of f_{inf}, the compressive prestress at the bottom and f_{sup}, the tensile prestress at the top, may be calculated from equations (4.1)–(4.4). Generally the minimum value of the compressive prestress and the maximum value of the tensile prestress in the concrete will be of most interest because the prestressing force, and hence the area of the tendons required will then be least. In certain circumstances, however, it may be necessary to select a prestress greater than the minimum.

Having decided on the prestress in the concrete, the next step is to calculate the theoretical eccentricity of the prestressing force. This may be done by means of the following formula derived from equations (2.2) and (2.3) in Chapter 2:

$$e = \frac{i^2(f_{inf}+f_{sup})}{h\left[f_{inf}-\dfrac{y_{inf}}{h}(f_{inf}+f_{sup})\right]} \tag{4.7}$$

If M_{min} is relatively large the value of e may be so great as to require the tendons to be located below the soffit of the beam, or too near the bottom of the section to allow sufficient cover. In this event the largest practicable value of e must be adopted.

The prestressing force, acting at an eccentricity e, must develop a prestress equal to f_{inf} in the concrete at the bottom of the section and is, therefore, given by:

$$P = \frac{f_{inf}A}{\left(1+\dfrac{y_{inf}e}{i^2}\right)} \tag{4.8}$$

The required area of the tendons can now be obtained, since the stress in the tendons at transfer is known approximately or may be calculated from the initial jacking stress as follows:

$$f_{pp} = f_{pi}-\alpha f_{cp} \tag{4.9}$$

where f_{pp} = stress in prestressing tendons at transfer of prestress
  f_{pi} = initial stress in prestressing tendons at time of jacking
  f_{cp} = stress in concrete at level of centroid of tendons, at transfer of prestress
  α = modular ratio E_s/E_c

If it has been necessary to reduce the value of e for practical reasons, the effect will be to reduce the tensile prestress f_{sup} at the top as indicated in Figure 4.5. Under the design load the resultant compressive stress at the top will therefore be increased and it will be necessary to check that it does not exceed the allowable stress $f_{c\ adm}$. This is one of the reasons why it is advisable to select a section with a value of Z_{sup} greater than the minimum given by formula (4.6).

Some of the practical considerations governing the arrangement of tendons will be demonstrated in continuing the previous example.

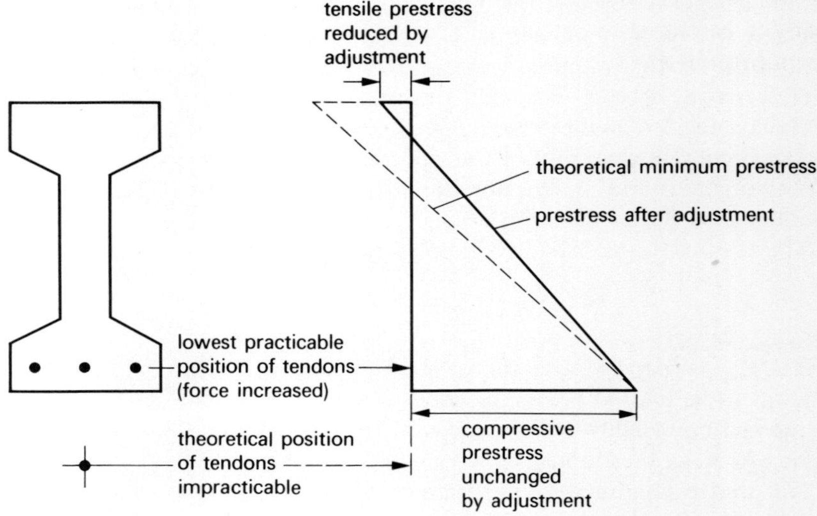

Figure 4.5 Modification of prestress due to adjustment of prestressing force

Example

(a) *Tendons parallel to axis*
For section DT3: $Z_{inf} = 8{\cdot}95 \times 10^6$ mm³
From formulae (4.3) and (4.2), minimum prestress in concrete:

$$f_{inf} = \frac{1}{0{\cdot}75}\left(\frac{101{\cdot}2}{8{\cdot}95} - 3{\cdot}2\right) = 10{\cdot}8 \text{ N/mm}^2 \text{ (compressive)}$$

$$f_{sup} = 2{\cdot}7 + 0 = 2{\cdot}7 \text{ N/mm}^2 \text{ (tensile)}$$

For section DT3:

$$h = 406 \text{ mm} \quad A = 157\,000 \text{ mm}^2 \quad y_{inf} = 264 \text{ mm}$$

$$i^2 = \frac{2365 \times 10^6}{157\,000} = 15\,060 \text{ mm}^2 \quad y_{inf}/h = 264/406 = 0{\cdot}650$$

From formula (4·7)

$$e = \frac{15\,060(10{\cdot}8 + 2{\cdot}7)}{406[10{\cdot}8 - 0{\cdot}650(10{\cdot}8 + 2{\cdot}7)]} = 248 \text{ mm}$$

This is a little too large to allow sufficient bottom cover. According to CP 110, 3.11.2, (Table 2.7), the minimum cover for Grade 50 under protected conditions is 15 mm, and a slightly reduced value of $e = 240$ mm will give

18·5 mm cover over strands of nominal diameter 10·9 mm, if there are not too many to be placed in one layer.

From formula (4.8)

$$P = \frac{10 \cdot 8 \times 157\,000}{(1 + 264 \times 240/15\,060)1000} = 326 \text{ kN}$$

If the tensile strength of the strand is 1770 N/mm², jacking stress in tendons $= 0.7 \times 1770 = 1240$ N/mm².

The stress at transfer will be approximately

$$0.9 \times 1240 = 1110 \text{ N/mm}^2$$

More accurately the compressive prestress in concrete at level of tendons

$$= \frac{326\,000}{157\,000}(1 + 240^2/15\,060) = 10 \cdot 0 \text{ N/mm}^2$$

From CP 110 2.4.2.4 and Table 1 (Tables 2.1 and 2.2), interpolating to obtain the modulus of elasticity corresponding to the concrete strength of 35 N/mm² at transfer:

$$E_s = 200 \text{ kN/mm}^2 \quad \text{and} \quad E_c = 29 \cdot 5 \text{ kN/mm}^2$$

Hence from formula (4.9) stress in steel at transfer is

$$f_{pp} = 1240 - \frac{200}{29 \cdot 5} \times 10 \cdot 0 = 1170 \text{ N/mm}^2$$

Area of tendons $= 326\,000/1170 = 279$ mm²

$4 - 10 \cdot 9$ mm tendons will be used having an area of 284 mm²

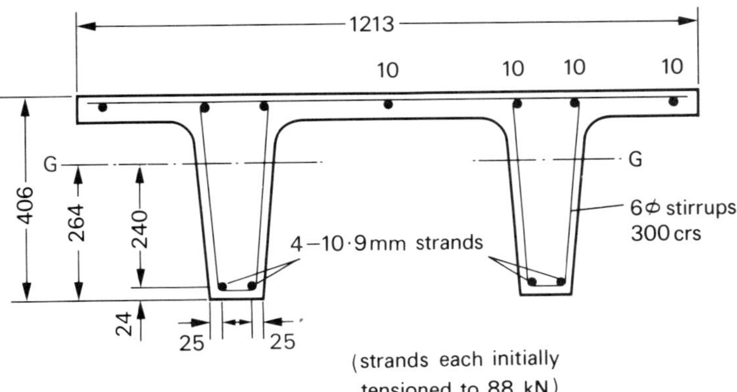

Figure 4.6 Arrangement of tendons in double-T section

The arrangement of the tendons in the section is shown in Figure 4.6. It should be borne in mind that the tendons have to provide the major component of tensile resistance at the ultimate limit state and as large a group of them as possible should therefore be located near the bottom of the section where they will also serve as reinforcement to control cracking in the event of an accidental overload. In our present example of a double-T section, which characteristically requires a prestressing force at a large eccentricity, this requirement is fulfilled. In beams of rectangular or I section, however, the required eccentricity will be less, and rather than place all the tendons nearer to the centroid of the section it is better to locate the majority near the bottom with a few tendons near the top face, so that the resultant prestressing force (i.e. the centre of area of all the tendons) is at the correct eccentricity. A further advantage of this arrangement, an example of which is illustrated in Figure 4.7, is that the top tendons can function as reinforcement to control accidental cracking arising from excessive shrinkage or from a reversal of loading. In the absence of tendons at the top this purpose must be achieved by the provision of reinforcement.

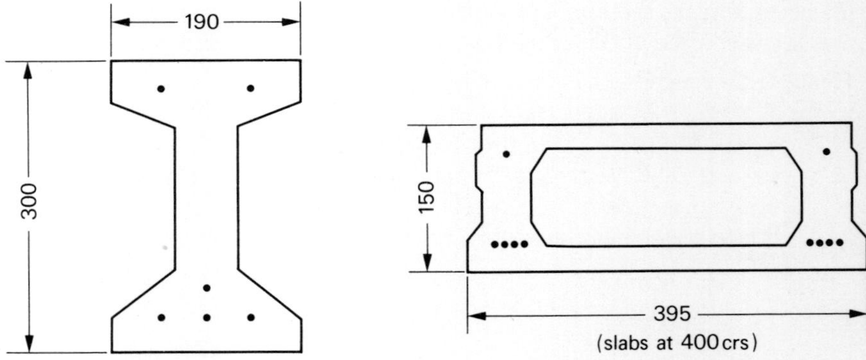

Figure 4.7 Typical arrangement of tendons in small I and box sections
(Concrete Ltd)

(b) *Deflected or debonded tendons*
The self-weight and total design load are reduced by the use of the smaller section DT2 so that:

$$M_{min} = 35 \cdot 6 \text{ kNm}; \quad M_d = 95 \cdot 7 \text{ kNm}$$

From formulae (4.3) and (4.2) minimum prestress in concrete

$$f_{inf} = \frac{1}{0 \cdot 75} \left(\frac{95 \cdot 7}{5 \cdot 57} - 3 \cdot 2 \right) = 18 \cdot 6 \text{ N/mm}^2 \text{ (compressive)}$$

$$f_{sup} = \frac{35\cdot6}{10\cdot66} + 2\cdot7 = 6\cdot0 \text{ N/mm}^2 \text{ (tensile)}$$

It will be observed that the values of the prestress f_{inf} and f_{sup} are both greater than the permissible stresses; this is of no practical significance, however, since bending stresses will always be present so that the *resultant* stresses will be within the permissible values.

For section DT2:

$$h = 305 \text{ mm} \quad A = 134\,000 \text{ mm}^2 \quad y_{inf} = 200 \text{ mm}$$

$$i^2 = \frac{1115 \times 10^6}{134\,000} = 8330 \text{ mm}^2 \qquad y_{inf}/h = 200/305 = 0\cdot657$$

From formula (4.7)

$$e = \frac{8330(18\cdot65 + 6\cdot04)}{305[18\cdot65 - 0\cdot657(18\cdot65 + 6\cdot04)]} = 276 \text{ mm}$$

This is obviously impracticable, as it would require the tendons to be below the bottom of the section. Estimating that the lowest possible position of the prestressing force is 35 mm from the bottom, e is therefore, adjusted to $200 - 35 = 165$ mm.

From formula (4.8)

$$P = \frac{18\cdot65 \times 134\,000}{(1 + 200 \times 165/8330)1000} = 503 \text{ kN}$$

compressive prestress in concrete at level of tendons

$$= \frac{492\,000}{134\,000}(1 + 170^2/8330) = 16\cdot3 \text{ N/mm}^2$$

steel stress at transfer, $f_{pp} = 1240 - \dfrac{200}{29\cdot5} \times 16\cdot35 = 1130 \text{ N/mm}^2$

area of tendons $= 503\,000/1130 = 445 \text{ mm}^2$

6–10·9 mm tendons will be used, having an area of 426 mm². A very slight increase must be made in the jacking stress originally assumed; the required prestressing may be obtained by an initial force of 92 kN (73 % of the tensile strength) on each strand.

The tendons are arranged as in Figure 4.8 section B-B so that the two lower strands in each rib can be continued parallel to the axis of the beam over the full length, and the upper strand can be deflected upwards at the ends. The exact eccentricity at the mid-span section is 167 mm.

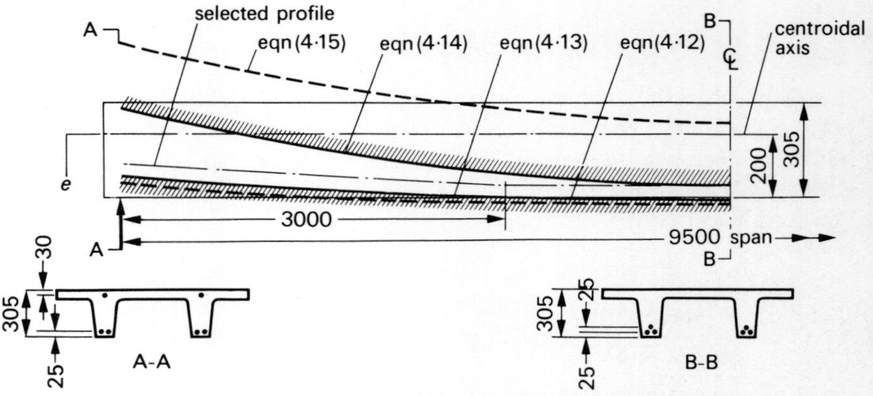

Figure 4.8 Arrangement of deflected tendons

Since there has been a large adjustment to the eccentricity, it is necessary to check the stress at the top of the section.

$$\text{Tensile prestress} = \frac{503\,000}{134\,000}\left(1 - \frac{105 \times 167}{8330}\right) = 4\cdot1 \text{ N/mm}^2$$

As previously, the resultant tensile stress when the minimum load is present will be less than the permissible value.

$$\text{Compressive stress at design load} = \frac{95\cdot5}{10\cdot66} - 4\cdot1 = 4\cdot8 \text{ N/mm}^2$$

This is well within the allowable stress of $16\cdot7$ N/mm².

4.5 DESIGN OF TENDONS FOR SERVICEABILITY AT REMAINING SECTIONS

When the tendons have been designed only for the critical section the next step is to arrange their profile so as to meet the stress conditions for serviceability over the whole of the member. This may be done by the use of deflected or debonded tendons. If the tendons are to be deflected the prestressing force P will be constant and its eccentricity e will vary along the beam. With debonded tendons, however, e will be more or less constant and P can be reduced by rendering some of the tendons ineffective over a length of the beam adjacent to each end. In many instances of either type a suitable arrangement may easily be found by trial and error; the direct method will, however, be described.

The limits within which the prestressing force must lie are given by four equations developed from the four stress conditions (4.1) to (4.4). The prestress formulae may be written

$$f_{\mathrm{inf}} = \frac{P}{A} + \frac{Pe}{Z_{\mathrm{inf}}} \quad \text{(compressive)} \qquad (4.10)$$

$$f_{\mathrm{sup}} = \frac{Pe}{Z_{\mathrm{sup}}} - \frac{P}{A} \quad \text{(tensile)} \qquad (4.11)$$

Substituting these values in equations (4.1) to (4.4)

$$e \leqslant \frac{Z_{\mathrm{inf}} f_{\mathrm{cp\,adm}}}{P} - \frac{Z_{\mathrm{inf}}}{A} + \frac{M_{\mathrm{min}}}{P} \qquad (4.12)$$

$$e \leqslant \frac{Z_{\mathrm{sup}} f_{\mathrm{tp\,adm}}}{P} + \frac{Z_{\mathrm{sup}}}{A} + \frac{M_{\mathrm{min}}}{P} \qquad (4.13)$$

$$e \geqslant -\frac{Z_{\mathrm{inf}} f_{\mathrm{t\,adm}}}{\eta P} - \frac{Z_{\mathrm{inf}}}{A} + \frac{M_{\mathrm{d}}}{\eta P} \qquad (4.14)$$

$$e \geqslant -\frac{Z_{\mathrm{sup}} f_{\mathrm{c\,adm}}}{\eta P} + \frac{Z_{\mathrm{sup}}}{A} + \frac{M_{\mathrm{d}}^{*}}{\eta P} \qquad (4.15)$$

Obviously only two of the above four conditions will be critical at any point on the beam, and if the beam is uniform in cross-section (i.e. prismatic) so that A, Z_{inf} and Z_{sup} are constant, equations (4.12) and (4.13) will be represented by two parallel curves on the longitudinal elevation, one of which defines the lowest possible profile of the prestressing force, while (4.14) and (4.15) will also be parallel curves, one of which is the upper limit of the prestressing force. This is shown in the example below.

Similarly equations (4.12) to (4.15) can be arranged to indicate the maximum and minimum possible values of the prestressing force when the eccentricity is fixed.

$$P \leqslant \frac{A(M_{\mathrm{min}}/Z_{\mathrm{inf}} + f_{\mathrm{cp\,adm}})}{A e/Z_{\mathrm{inf}} + 1} \qquad (4.16)$$

$$P \leqslant \frac{A(M_{\mathrm{min}}/Z_{\mathrm{sup}} + f_{\mathrm{tp\,adm}})}{A e/Z_{\mathrm{sup}} - 1} \qquad (4.17)$$

$$P \geqslant \frac{A(M_{\mathrm{d}}/Z_{\mathrm{inf}} - f_{\mathrm{t\,adm}})}{\eta(A e/Z_{\mathrm{inf}} + 1)} \qquad (4.18)$$

$$P \geqslant \frac{A(M_{\mathrm{d}}/Z_{\mathrm{sup}} - f_{\mathrm{c\,adm}})}{\eta(A e/Z_{\mathrm{sup}} - 1)} \qquad (4.19)$$

Example

The design of the double-T section DT2 is continued.

---

* Strictly, here and in formula (4.19) the loss ratio η should be put equal to unity when $e \leqslant Z_{\mathrm{sup}}/A$.

(a) *Deflected tendons*

The boundaries within which the prestressing force must lie are established by inserting the appropriate numerical values into formulae (4.12) to (4.15).

Formula (4.12)

$$e \leqslant \frac{5 \cdot 57 \times 17 \cdot 5}{0 \cdot 503} - \frac{5 \cdot 57}{0 \cdot 134} + \frac{M_{min}}{0 \cdot 503} = 151 + \frac{M_{min}}{0 \cdot 503} \text{ mm}$$

Formula (4.13)

$$e \leqslant \frac{10 \cdot 66 \times 2 \cdot 2}{\cdot 0 \cdot 503} + \frac{10 \cdot 66}{0 \cdot 134} + \frac{M_{min}}{0 \cdot 503} = 126 + \frac{M_{min}}{0 \cdot 503} \text{ mm}$$

Formula (4.14)

$$e \geqslant -\frac{5 \cdot 57 \times 3 \cdot 2}{0 \cdot 75 \times 0 \cdot 503} - \frac{5 \cdot 57}{0 \cdot 134} + \frac{M_d}{0 \cdot 75 \times 0 \cdot 503} = -89 + \frac{M_d}{0 \cdot 378} \text{ mm}$$

Formula (4.15)

$$e \geqslant -\frac{10 \cdot 66 \times 16 \cdot 7}{0 \cdot 503} + \frac{10 \cdot 66}{0 \cdot 134} + \frac{M_d}{0 \cdot 503} = -293 + \frac{M_d}{0 \cdot 503} \text{ mm}$$

These are drawn in Figure 4.8, and it is seen that the zone within which the prestressing force must lie is defined by formulae (4.13) and (4.14). These conditions may be satisfied by deflecting the two upper tendons from points 3 m from the supports, so as to be 30 mm from the top over the supports, as shown in Section A-A.

(b) *Debonded tendons*

The limiting values of the prestressing force are indicated by formulae (4.16) to (4.19).

Formula (4.16)

$$P \leqslant \frac{134(M_{min}/5 \cdot 57 + 17 \cdot 5)}{0 \cdot 134 \times 167/5 \cdot 57 + 1} = 4 \cdot 80 M_{min} + 468 \text{ kN}$$

Formula (4.17)

$$P \leqslant \frac{134(M_{min}/10 \cdot 66 + 2 \cdot 7)}{0 \cdot 134 \times 167/10 \cdot 66 - 1} = 11 \cdot 43 M_{min} + 329 \text{ kN}$$

Formula (4.18)

$$P \geqslant \frac{134(M_d/5 \cdot 57 - 3 \cdot 2)}{0 \cdot 75(0 \cdot 134 \times 167/5 \cdot 57 + 1)} = 6 \cdot 40 M_d - 114 \text{ kN}$$

Formula (4.19)

$$P \geqslant \frac{134(M_d/10 \cdot 66 - 16 \cdot 7)}{0 \cdot 75(0 \cdot 134 \times 167/10 \cdot 66 - 1)} = 15 \cdot 23 M_d - 2710 \text{ kN}$$

Formulae (4.16) to (4.18) enable a diagram to be drawn as in Figure 4.9, showing the range of possible values of P at each point on the beam. In this example formula (4.19) yields negative values of P and is therefore not relevant.

The conditions may be fulfilled by debonding the end lengths of the two upper tendons (Figure 4.8, Section B-B) so that they are not fully effective at a distance less than 1·8 m from each support. Debonding is usually effected by placing a sleeve over the length of tendon which is not required. The sleeve should be sufficiently rigid to withstand the pressure of the fresh concrete without deforming and gripping the tendon.

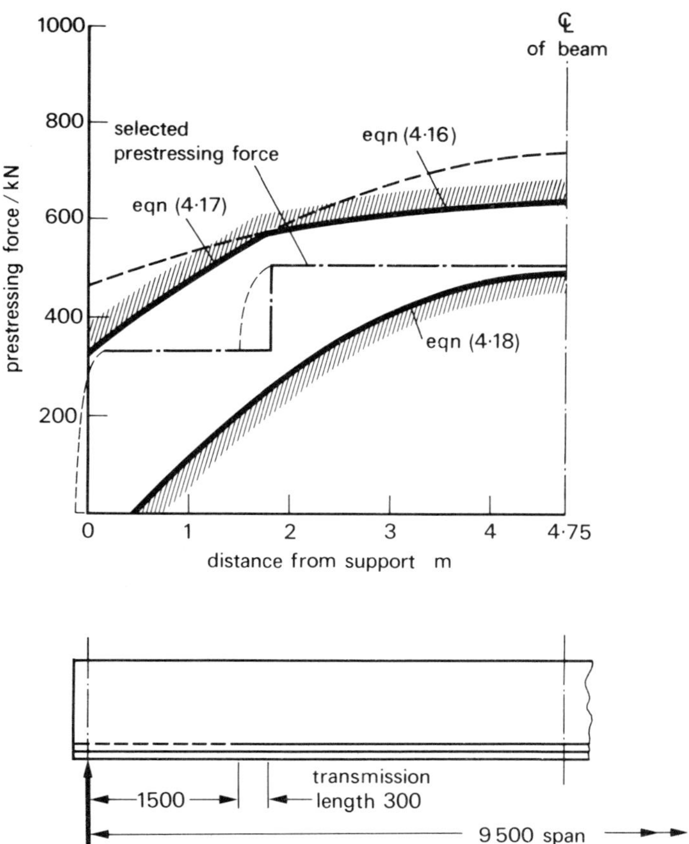

Figure 4.9 Arrangement of debonded tendons

4.6 TRANSMISSION LENGTH OF TENDONS

The transmission length of tendons has been discussed in Chapter 2, and it will be recalled that the stress in each pretensioned wire or strand increases from zero at the free end to the transfer stress at the end of the transmission length. The increase of stress is not linear but occurs at a greater rate near the end, so that 80% of the transfer stress may be developed in about 70% of the transmission length.

The fact that the prestress is not fully effective within the transmission length presents few problems in simply supported beams unless a load is applied very close to a support. The transmission length must, however, be taken into account when calculating the position at which the debonding of tendons may commence.

Example

The transmission length of 10·9 mm strand used in the previous example is found from CP 110, Table 42, by interpolation, to be about 265 ± 25 mm. If the overall length of the beam is 9750 mm (250 mm greater than the span) the prestress will be fully effective at a maximum distance of $265 \pm 25 - 125 = 165$ mm from the support point at each end.

The development of the prestressing force in the debonded tendons is shown by dotted lines in Figure 4.9. Debonding of the tendons should, therefore, commence at a distance of $1800 - 300 = 1500$ mm from the support, or 1625 mm from the end of the beam.

4.7 CHECK OF SECTIONS FOR ULTIMATE MOMENT

It is often sufficient to check the ultimate moment of a beam at one critical section, normally the section at mid-span. When, however, the cross-section varies or the tendons are deflected it may be necessary to calculate the ultimate moment at several other sections, and perhaps to compare the diagrams of ultimate moment of resistance and the required ultimate moment (i.e. the design ultimate moment) to ensure that the ultimate limit state conditions are satisfied at all points along the beam.

The design ultimate load is computed by applying the factors specified in CP 110 2.3.3.1 (Table 1.1) to the characteristic dead, live and, where applicable, wind load. For American practice the corresponding factors are given in ACI 318–71, 9.3.

The principles for the analysis of sections under ultimate loads are given in CP 110 4.3.4 and have been explained in Chapter 3. The complete analysis according to the assumptions in CP 110 4.3.4.1 is somewhat laborious, and the simpler method, here described, will be found to be sufficiently accurate in almost all cases.

The procedure is as follows:

(1) Calculate the depth of the compression zone and the required compression resistance of the concrete.

(a) Rectangular section

The stress distribution in the concrete is shown in Figure 4.10, assuming the stress to be 0·4 times the characteristic cube strength over the whole compression zone. This would be approximately equivalent to 0·6 times the cylinder strength when working to ACI 318–71.

If M_{ud} is the design ultimate moment:

$$M_{ud} = 0{\cdot}4f_{cu}bx(d-0{\cdot}5x) \tag{4.20}$$

Hence $$x/d = 1-[1-5(M_{ud}/f_{cu}bd^2)]^{\frac{1}{2}} \tag{4.21}$$

x/d should not be greater than the maximum value provided for in CP 110, Table 37, i.e. 0·783 for pretensioning

then $$N_{cu} = 0{\cdot}4f_{cu}bx \tag{4.22}$$

(b) I, T or box sections

The stress is again assumed to be $0{\cdot}4f_{cu}$ over the whole compression zone, and the centre of compression is assumed to be at the centroid of the area of concrete in compression.

When the compression zone lies completely within the top flange ($x \leqslant d_f$) the procedure is the same as for a rectangular section. When the compression zone includes the upper part of the web an approximate method is to neglect this area, so that

$$x = d_f$$

$$N_{cu} = 0{\cdot}4f_{cu}bd_f \tag{4.23}$$

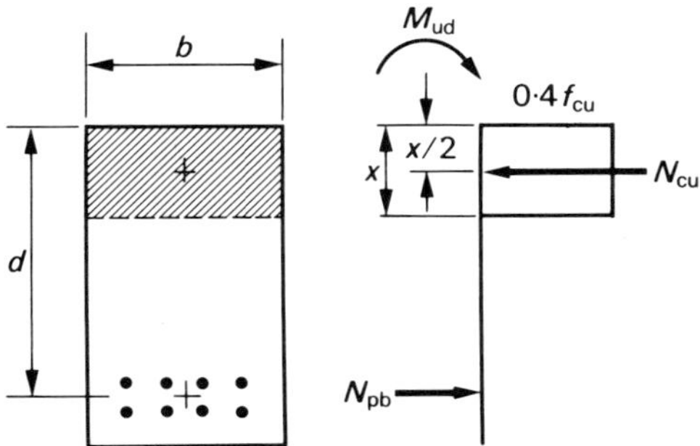

Figure 4.10 Ultimate limit state of rectangular section in bending

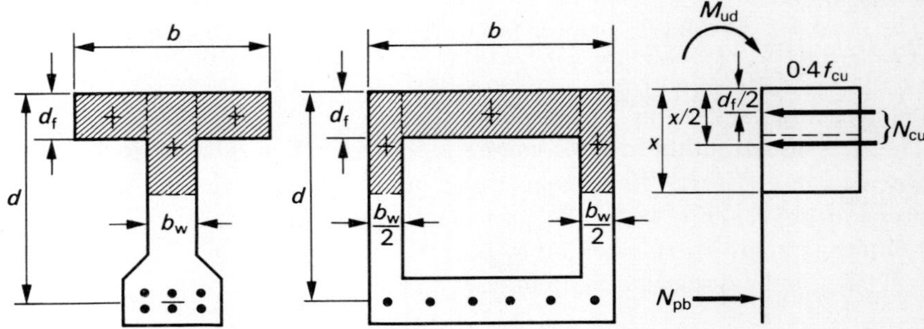

Figure 4.11 Ultimate limit state of non-rectangular sections in bending

Alternatively, referring to Figure 4.11

$$M_{ud} = 0 \cdot 4 f_{cu}[(b - b_w)d_f(d - 0 \cdot 5 d_f) + b_w x(d - 0 \cdot 5 x)]$$

Hence
$$x/d = 1 - [1 - 5(M_{wu}/f_{cu}b_{uw}d^2)]^{\frac{1}{2}} \tag{4.24}$$

where
$$M_{wu} = M_{ud} - 0 \cdot 2 f_{cu}(b - b_w)d_f(2d - d_f)$$

then
$$N_{cu} = 0 \cdot 4 f_{cu}[(b - b_w)d_f + b_w x] \tag{4.25}$$

(2) Obtain the value of f_{pb}, the stress in the tendons, from the relationship between x/d and $f_{pb}/0 \cdot 87 f_{pu}$ in CP 110, Table 37, as shown in Figure 3.6, and hence the tensile resistance of the tendons.

$$N_{pb} = f_{pb}A_p \tag{4.26}$$

If $N_{pb} \geqslant N_{cu}$ no additional reinforcement is necessary; otherwise A_s the required area of reinforcement of yield stress f_y is given by:

$$A_s = \frac{N_{cu} - N_{pb}}{0 \cdot 87 f_y} \tag{4.27}$$

This assumes that the tensile strain in the reinforcement is sufficient to develop the yield stress, as will be so if the bars are located near the bottom.

Occasionally the maximum value of x may not result in a sufficiently large value of N_{cu} to balance the design ultimate moment, and compressive reinforcement may be required. More commonly there will be a few pretensioned tendons in the compressive zone but, except where the value of x is large, these will not cause a significant change in the ultimate moment, as the force exerted by them will be balanced by an increase in the size of the compression zone. This calculation has been more fully discussed elsewhere ref. (4.3).

Example.

The double-T section DT3, designed with tendons parallel to the axis (Figure 4.6) will be checked for ultimate moment.

From CP 110 2.3.3.1 and specified loading (p. 4.8)

$$\text{design ultimate load} = 1\cdot4 \times 4\cdot4 + 1\cdot6 \times 3\cdot0 = 11\cdot0 \text{ kN/m}^2$$

$$M_u = 11\cdot0 \times 1\cdot213 \times 9\cdot5^2 \times 0\cdot125 = 150 \text{ kNm}$$

$$M_u/f_{cu}bd^2 = 150 \times 10^6/(50 \times 1213 \times 382^2) = 0\cdot017$$

From formula (4.21)

$$x/d = 1 - (1 - 5 \times 0\cdot017)^{\frac{1}{2}} = 0\cdot047$$

$$x = 0\cdot047 \times 382 = 18 \text{ mm}$$

Compression zone lies within top slab. Therefore from formula (4.22)

$$N_{cu} = 0\cdot4 \times 50 \times 1213 \times 18/1000 = 437 \text{ kN}$$

From CP 110, Table 37 or Figure 3.4, $f_{pb} = 0\cdot87 f_{pu}$
hence from formula (4.26)

$$N_{pb} = 4 \times 0\cdot87 \times 1770 \times 71/1000 = 438 \text{ kN}$$

This is about equal to N_{cu}, the force required to develop the design ultimate moment, and no additional reinforcement is required. As an alternative method, the required area of the tendons could be obtained from Column 1 of Table 37, but this column is not applicable when the compression zone is non-rectangular.

4.8 CHECK OF BEAMS FOR ULTIMATE SHEAR

The factors affecting the ultimate shear and the recommended formulae have been discussed in the previous chapter, from which it will be recalled that the mode of shear failure differs according to whether or not the beam is cracked in flexure at the time of failure. When carrying out a shear check of a member, such as a simply supported beam, in which the maximum shear does not coincide with a large moment, it will therefore be necessary to consider at least two points.

In CP 110 4.3.5.2 the ultimate shear resistance of a section cracked in flexure is related to the moment M_0 necessary to produce zero stress in the concrete, at the depth d and, in checking for shear, flexural cracking should be assumed to have occurred where the moment resulting from the design ultimate load is greater than M_0.

In a simply supported beam subject to a uniformly distributed load the moment is M at a distance x from the support, given by:

$$x = \frac{l}{2}\left[1-(1-(M/M_{max})^{\frac{1}{2}}\right] \tag{4.28}$$

The shear at this point is given by

$$V = V_{max}(1-2x/l) \tag{4.29}$$

The critical cracked section for shear will be at the point where $V_{cr} \sim V$ is a minmium. Substituting for V_{cr} from CP 110 4.3.5.2, formula (46) [given on p. 37 as formula (3.6)]:

$$\frac{d}{dx}\left[(1-0\cdot55f_{pe}/f_{pu})v_c bd+\frac{M_0 wx(l-2x)}{2wx(l-x)}-\frac{w(l-2x)}{2}\right] = 0$$

When the prestress and reinforcement are constant for all values of x, as in a beam with tendons parallel to the axis, the first term of the expression and M_0 are both constant. It can then be shown that:

$$\frac{8x^2(1-x)^2}{l^2(2x^2-2xl+l^2)} = \frac{M_0}{M_{max}} \tag{4.30}$$

The solution of this equation may be obtained for any given value of M_0/M_{max} from Figure 4.12. The shear strength is then checked according to CP 110, 4.3.5 or ACI 318–71, 11.5 and 11.6.

Example

The respective calculations for sections uncracked and cracked in flexure will be applied to carry out a shear check on the double-T beam DT3 which has been designed for flexure and checked for ultimate moment.

(a) Section at support (uncracked in flexure).

Ultimate shear $V_{ud} = 11\cdot0\times1\cdot213\times9\cdot5\times0\cdot5 = 63\cdot3$ kN

At centroid of section $b = 150\times2 = 300$ mm

$h = 406$ mm

Tensile strength of concrete $f_t = 0\cdot24(f_{cu})^{\frac{1}{2}} = 0\cdot24(50)^{\frac{1}{2}} = 1\cdot7$ N/mm^2

Prestress at centroid (after all losses)

$$= \frac{\eta P}{A} = \frac{0\cdot75\times284\times1170}{157\ 000} = 1\cdot6 \text{ N/mm}^2$$

From CP 110 4.3.5.1, formula (45) [p. 36 formula (3.5)]

$$V_{co} = 0\cdot67bh(f_t^2+0\cdot8f_{cp}f_t)^{\frac{1}{2}}$$

$$= 0\cdot67\times300\times406(1\cdot7^2+0\cdot8\times1\cdot6\times1\cdot7)^{\frac{1}{2}}$$

$$= 183 \text{ kN}$$

$$V_{co} > V_{ud}$$

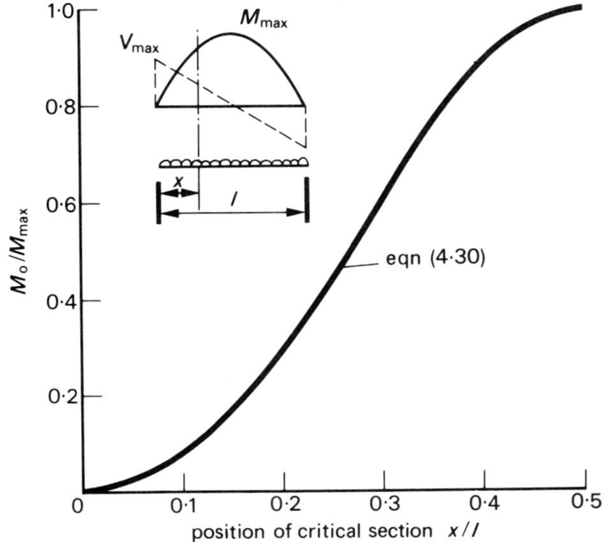

Figure 4.12 Position of critical shear section in a beam with uniform prestress, cracked in flexure under uniformly distributed load

(b) Section cracked in flexure with maximum shear ($M = M_0$).
At depth $d = 382$ mm (level of tendons), $y = 240$ mm.
Prestress in concrete

$$f_{pt} = \frac{0\cdot75 \times 284 \times 1170}{157\ 000}\left(1 + \frac{240^2}{15\ 060}\right) = 7\cdot7\ \text{N/mm}^2$$

$$M_0 = 0\cdot8 f_{pt}\frac{I}{y} = 0\cdot8 \times 7\cdot7 \times \frac{2365 \times 10^6}{240} \times \frac{1}{10^6} = 60\cdot5\ \text{kNm}$$

Formula (4.28) gives the point on the beam at which $M = M_0$

$$x = \frac{9\cdot50}{2}\left[1 - (1 - 60\cdot5/150\cdot0)^{\frac{1}{2}}\right] = 1\cdot08\ \text{m}$$

Since $M_0/M_{max} = 0\cdot403$, the position of the critical section is found from Figure 4.12 to be:

$$x = 0\cdot235 \times 9\cdot50 = 2\cdot23\ \text{m}$$

At this point:

$$M = \frac{wx}{2}(l - x) = 11\cdot0 \times 1\cdot213 \times 2\cdot23(9\cdot50 - 2\cdot23)/2 = 108\ \text{kNm}$$

and from formula (4.29)

$$V = 63 \cdot 3(1 - 2 \times 0 \cdot 235) = 33 \cdot 5 \text{ kN}$$

$$f_{pe} = 0 \cdot 75 \times 1170 = 877 \text{ N/mm}^2 \quad f_{pu} = 1770 \text{ N/mm}^2$$

$$\frac{100 A_p}{bd} = \frac{100 \times 284}{300 \times 382} = 0 \cdot 25\%$$

From CP 110, 3.3.6.1, Table 5 (Figure 3.8): $v_c = 0 \cdot 35 \text{ N/mm}^2$
From CP 110, 4.3.5.2, formula (46) [p. 37 formula (3.6)]

$$V_{cr} = \left(1 - 0 \cdot 55 \frac{f_{pe}}{f_{pu}}\right) v_c bd + M_0 \frac{V}{M}$$

$$= (1 - 0 \cdot 55 \times 877/1770)0 \cdot 35 \times 300 \times 382/1000 + 60 \cdot 5 \times 33 \cdot 5/107 \cdot 8$$

$$= 29 \cdot 2 + 18 \cdot 8 = 48 \cdot 0 \text{ kN}$$

$$V_{cr} > V_{ud} > V_{cr}/2$$

According to CP 110 4.3.5.3, minimum shear reinforcement must be provided where $V > V_{cr}/2$ unless the beam is considered to be 'of minor importance'. Using steel with a characteristic strength of 410 N/mm², the required area of reinforcement is given by

$$\frac{A_{sv}}{s_v} = \frac{0 \cdot 4b}{0 \cdot 87 f_{yv}} = \frac{0 \cdot 4 \times 300}{0 \cdot 87 \times 410} = 0 \cdot 336 \text{ mm}^2/\text{mm}$$

6 mm stirrups at 300 mm centres in each web, as in Figure 4.6, provide an area of 0·376 mm²/mm. Although they are not required where $V < V_{cr}/2$, it would probably be more convenient to continue them over the full length of the beam.

4.9 ECONOMY

The examples in this chapter have demonstrated alternative solutions to a design problem, the one using parallel tendons to produce a uniform prestress at all sections, and the other varying the prestress by deflecting or debonding the tendons. Design decisions are often based on economic considerations, and it must be emphasized that the least expensive solution is not necessarily obtained with the smallest section, or even with the minimum amount of material.

In the examples of the design of double-T beams the use of deflected or debonded tendons enabled a smaller section to be specified. This would result in a 13% saving in the quantity of concrete, and the corresponding reduction of dead weight might in certain circumstances lead to some economy in the supporting structure. However, this saving was only achieved at the expense of a 50% increase in the number of tendons, while the introduction of deflection or debonding would entail a further increase in labour and equipment costs. The slightly reduced cost of the smaller mould would be insignificant if spread over a large number of re-uses. It is therefore probable that the larger unit DT3 would be the more economical to produce.

References

4.1 EVANS, R. H. and BENNETT, E. W., *Prestressed Concrete Theory and Design*, Chapman and Hall, London, 1958, Chapter 5.
4.2 BENNETT, E. W., *Structural Concrete Elements*, Chapman and Hall, London, 1973, Chapter 7.
4.3 *Ibid.*, Chapter 4.

5

Beams with Post-Tensioned Tendons

5.1 PRESTRESSING TECHNIQUE

The post-tensioning process is illustrated in Figure 5.1, which is the counterpart of Figure 4.1 in the previous chapter. Prestressing does not take place until the concrete has hardened, but at the time of casting provision must be made to accommodate the tendons. This may be by means of longitudinal ducts through the member, by grooves at the side of the member, or by external spacing guides placed at intervals along the web of a member of I section or inside the walls of large box beams. Ducts are frequently formed by means of metal sheaths placed in the mould with the tendons already in position; in other cases the tendons are inserted after the concrete has set sufficiently to permit withdrawal of the mandrels or inflated tubes forming the ducts.

Prestressing is accomplished either by securing one end of the tendon (the 'dead end') to the member by a special anchorage and applying a tensile force to the other end by means of a portable jack, or alternatively by jacking at both ends simultaneously. At the time of stressing the reaction of the tendon at the jacking end acts on the end of the member through the jack itself, but on completion of the operation the tendon is anchored directly to the member, enabling the jack to be removed. Finally the tendons are usually protected and bonded to the concrete by injecting cement grout into the duct, or in the case of external tendons by the application of cement mortar.

Tendons for post-tensioning may consist of cables made up of a number of parallel wires or strands, or of single bars. Table 5.1 gives the main sizes in current use in Great Britain.

The function of the anchorages is to secure the ends of each wire, strand or

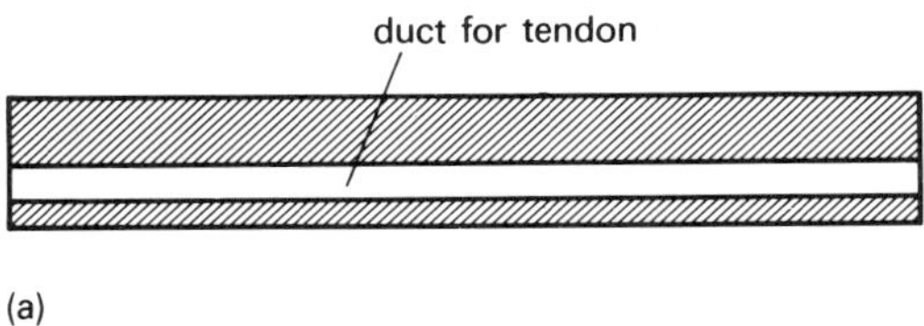

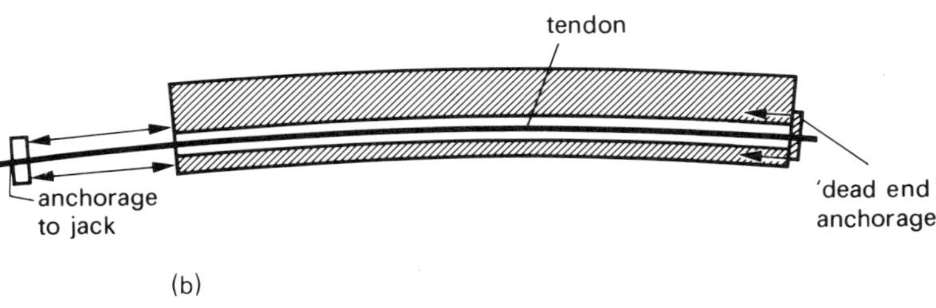

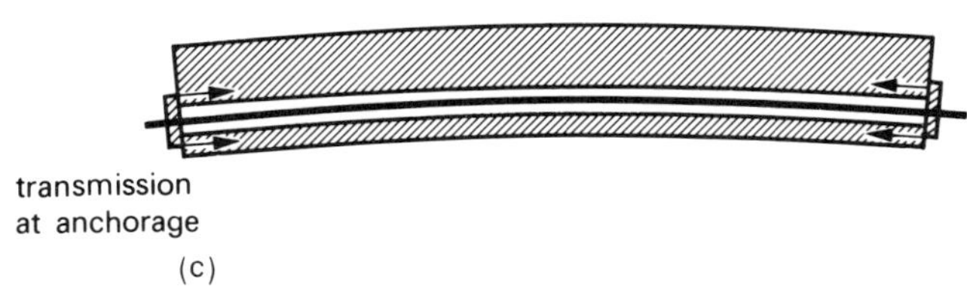

Figure 5.1 Post-tensioning process: (a) casting; (b) tensioning (transfer of prestress); (c) anchoring

bar, and to transmit the reaction to the end of the beam. Wedges are the commonest type of securing device, but a threaded nut may be used, particularly on bars, while in one system the securing of wires is effected by deforming the ends. The reaction may be transmitted to the concrete beam by a plate bearing on the end face, by an embedded cone or by a combination

Table 5.1 *Tendons for post-tensioning*

Wires (BS 2691 Section 2)		Strands (BS 3617 Section 2)		
Diameter/mm	*Characteristic tensile strength*/N/mm²	*Nominal diameter*/mm	*Characteristic breaking load*/kN	*Nominal steel area*/mm²
5·0	1570	7 *wire*		
	1670	6·4	44·5	24·5
	1720	7·9	69·0	37·4
6·0	1470	9·3	93·5	52·3
	1570	10·9	125·0	71·0
	1670	12·5	165·0	94·2
7·0		15·2	227·0	138·2
8·0	1470			
	1570	19-*wire*		
Bars (BS 4486)		18·0	370	210
		25·4	659	423
		28·6	823	535
Diameter/mm	*Characteristic breaking load*/N/mm			
20	325			
25	500			
32	800			
40	1250			

Note: Preferred sizes and strengths in bold figures.

of these two means. Several proprietary anchorage systems are available, current data for which should be obtained from the manufacturers. Problems of cover, spacing and reinforcement associated with anchorages will be considered later.

One of the main advantages of post-tensioning is the ease with which the line of the prestressing force may be varied by the use of flexible duct formers and tendons. A limitation is imposed, however, by the friction where the tendon bears on the wall of the duct or the spacing grilles, on the inside of curved lengths. The result is a progressive loss of stress, the calculation of which has been discussed in Chapter 2.

The anchorages required for post-tensioning are a relatively expensive feature which is absent from the pretensioning system. In addition the stressing operation itself calls for skilled labour and supervision on site. Consequently post-tensioning is now generally restricted to large members, cast and prestressed *in situ* and there is a trend toward larger sizes of tendon. However, the method opens up the possibility of new techniques, such as, for example, the connection of precast elements by means of post-tensioned tendons, and 'cantilevering out' in bridge construction.

5.2 DIMENSIONING OF CONCRETE SECTION FOR SERVICEABILITY

The general method used for beams with pretensioned tendons in the previous chapter is equally applicable to beams prestressed by post-tensioned tendons, and there are only minor differences arising from the different arrangement of the tendons in the section. An example will again be used to illustrate the procedure and some of the practical points.

Example

Design of Roof Girder

A simply supported roof girder with a span of 25 m is to be designed for a dead load of 2 kN/m and an imposed load of 15 kN/m in addition to the self-weight.

Moments at mid-span:

$$\text{self-weight (est. 12 kN/m)} \quad M_{\min} = 12 \times 25^2 \times 0.125 \quad = \quad 937 \text{ kNm}$$

$$\text{total dead load} \quad M_g = (12+2) \times 25^2 \times 0.125 = 1094 \text{ kNm}$$
$$\text{imposed load} \quad M_q = 15 \times 25^2 \times 0.125 \quad = 1172 \text{ kNm}$$

$$\text{design load} \quad M_d = M_g + M_q \quad = 2266 \text{ kNm}$$
(serviceability limit state)

Specifying concrete of Grade 40, to have a strength of 35 N/mm² at the time of prestressing and 40 N/mm² when the design load is first applied, the allowable stresses according to CP 110 are:

compressive stress at transfer (Table 36 or Table 2.5)
$$f_{cp\,adm} = 0.5 \times 35 = 17.5 \text{ N/mm}^2$$

tensile stress at design load (Table 33 or Table 2.6)
$$f_{t\,adm} = 2.3 \text{ N/mm}^2$$

The loss ratio η is assumed to be 0.80.

Required section modulus [formula (4.5)]

$$Z_{\inf} \geqslant \frac{M_d - \eta M_{\min}}{\eta f_{cp\,adm} + f_{t\,adm}}$$

$$= \frac{(2266 - 0.8 \times 937) \times 10^6}{0.8 \times 17.5 + 2.3} = 93.0 \times 10^6 \text{ mm}^3$$

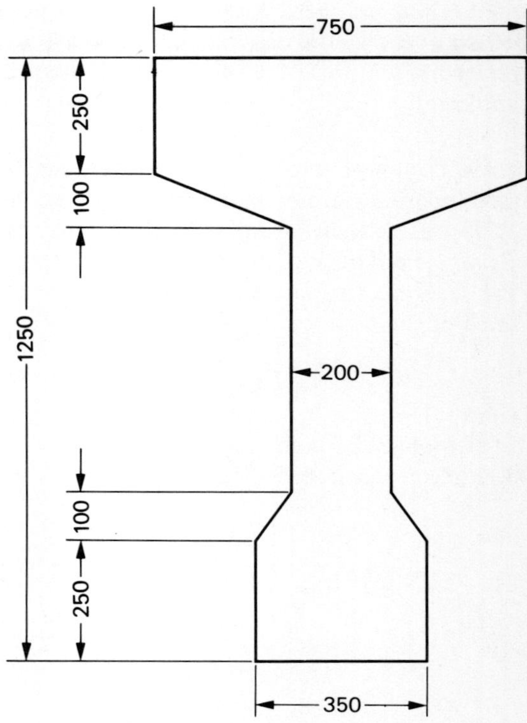

Figure 5.2 Dimensions of concrete section

In the absence of standard beam sections it is necessary to design a cross-section having a section modulus greater than the above value. A possible solution based on a depth/span ratio of 1/20 is shown in Figure 5.2; the dimensional properties of this section are as follows:

$$\text{area } A = 460\,000 \text{ mm}^2 \text{ (self-weight} \simeq 10\cdot8 \text{ kN/m)}$$

height of centroid $\qquad y_{\text{inf}} = 748$ mm

second moment of area $\qquad I = 74\,260 \times 10^6$ mm^4

radius of gyration $\qquad i^2 = I/A = 161\,500$ mm^2

section modulus $\qquad Z_{\text{inf}} = \dfrac{I}{y_{\text{inf}}} = 99\cdot3 \times 10^6$ mm^3

$$Z_{\text{sup}} = \dfrac{I}{y_{\text{sup}}} = 147\cdot8 \times 10^6 \text{ mm}^3$$

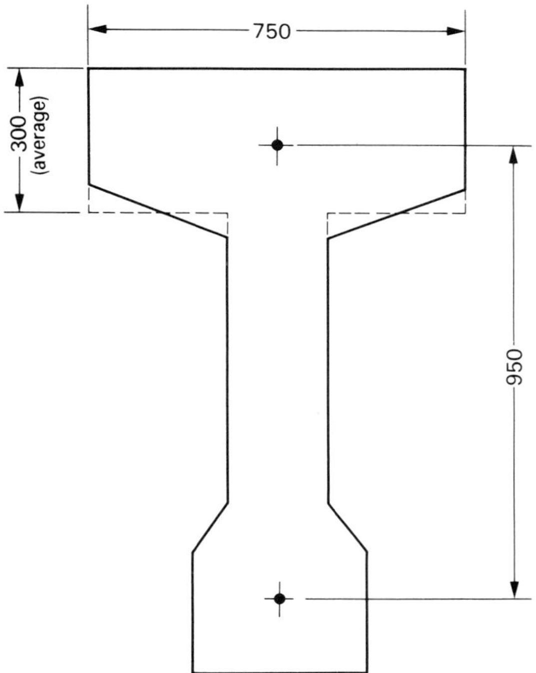

Figure 5.3 Check of area of top flange

Z_{sup} is well in excess of the required value, which will be found from formula (4.6) to be $100 \cdot 5 \times 10^6$ mm^3. This is because the top flange must be sufficient to provide the necessary moment of resistance of the concrete in compression at the ultimate limit state, which at this stage it is desirable to check as follows

$$\text{design ultimate moment } M_{ud} = 1 \cdot 4 M_g + 1 \cdot 6 M_q$$

$$= 1 \cdot 4 \times 1094 + 1 \cdot 6 \times 1172$$

$$= 3407 \text{ kNm}$$

average depth of top flange (Figure 5.3) $= 300$ mm

assuming centre of steel is located at centre of bottom flange,

$$\text{lever arm } z = 1250 - 300/2 - 300/2 = 950 \text{ mm}$$

ultimate moment of resistance

$$M_u \simeq 0 \cdot 4 \times 40 \times 750 \times 300 \times 950/10^6 = 3420 \text{ kNm}$$

D

The breadth of the web must be sufficient to accommodate tendons which are curved upwards at the ends of the beam; in addition, too thin a web may result in later difficulties when the shear resistance is checked at the ultimate limit state.

Before proceeding further, the deflection may, if desired, be checked at the serviceability limit state.

Short-term modulus of elasticity of concrete (CP 110, Table 1 or Table 2.1)

$$= 31 \text{ kN/mm}^2$$

Specific creep of concrete (CP 110, 4.8.2.5 or Table 2.3)

$$= 36 \times \frac{40}{35} \times 10^{-6} = 41 \times 10^{-6} \text{ per N/mm}^2$$

Long term modulus of elasticity $= \dfrac{31}{1+41 \times 10^{-6} \times 31 \times 10^3} = 13{\cdot}7 \text{ kN/mm}^2$

Using long term modulus for dead load and short-term modulus for imposed load

$$\text{deflection} = \frac{5 \times 14 \times 24^4 \times 10^9}{384 \times 13{\cdot}7 \times 74\,260 \times 10^6} + \frac{5 \times 15 \times 25^4 \times 10^9}{384 \times 31 \times 74\,260 \times 10^6}$$

$$= 70 + 33 \qquad\qquad = 103 \text{ mm}$$

This is 1/243 of the span, but the initial upward deflection due to the prestress has to be deducted. There will, therefore, be no difficulty in meeting the requirement of CP 110, 2.2.3.1(2), namely a final deflection not exceeding span/250 below the level of the supports.

5.3 DESIGN OF TENDONS

As in the earlier example, the minimum prestress required in the concrete at the midspan section is given by formulae (4.3) and (4.2)

$$f_{\text{inf}} = \frac{1}{\eta}\left(\frac{M_{\text{d}}}{Z_{\text{inf}}} - f_{\text{adm}}\right)$$

$$= \frac{1}{0{\cdot}8}\left(\frac{2266}{99{\cdot}3} - 2{\cdot}3\right) = 25{\cdot}6 \text{ N/mm}^2 \text{ (compression)}$$

$$f_{\text{sup}} = \frac{M_{\text{min}}}{Z_{\text{sup}}} + f_{\text{tp adm}}$$

$$= \frac{937}{147{\cdot}8} + 2{\cdot}2 = 8{\cdot}5 \text{ N/mm}^2 \text{ (tension)}$$

The theoretical eccentricity of the prestressing force is obtained from formula (4.7)

$$e = \frac{i^2(f_{inf}+f_{sup})}{h\left[f_{inf}-\dfrac{y_{inf}}{h}(f_{inf}+f_{sup})\right]}$$

$$= \frac{161\ 500(25{\cdot}6+8{\cdot}5)}{1250\left[25{\cdot}6-\dfrac{748}{1250}(25{\cdot}6+8{\cdot}5)\right]} = 848\ \text{mm}$$

This would require the prestressing force to act at a point below the bottom of the section. Assuming the lowest practical position of the centre of the tendons to be 200 mm above the bottom, e is adjusted to

$$748-200 = 548\ \text{mm}$$

$$\text{prestressing force } P = \frac{Af_{inf}}{1+ey_{inf}/i^2}$$

$$= \frac{460\ 000 \times 25{\cdot}6}{(1+548 \times 748/161\ 500)\ 1000} = 3328\ \text{kN}$$

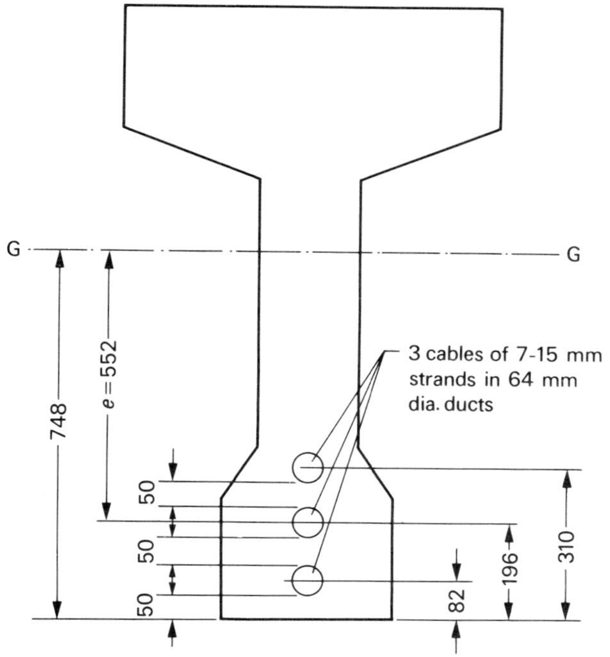

Figure 5.4 Position of tendons at mid-span section

The required force will be provided by 3 cables of 7–15 mm strands, each cable tensioned to 1110 kN. If they are arranged as in Figure 5.4, allowing 50 mm spacing and cover so as to permit free passage of the concrete when casting, the exact value of the eccentricity of the prestressing force will be

$$748 - 196 = 552 \text{ mm}$$

The stresses at the top of the section are now checked:
prestress at transfer

$$f_{\text{sup}} = \frac{3\,330\,000}{460\,000} \left(\frac{552 \times 502}{161\,500} - 1 \right) = 5 \cdot 2 \text{ N/mm}^2 \text{ (tension)}$$

compressive stress under design load

$$= \frac{2266}{147 \cdot 8} - 0 \cdot 8 \times 5 \cdot 2 = 11 \cdot 2 \text{ N/mm}^2$$

This is acceptable, since the allowable compressive stress $f_{c\,\text{adm}}$ for Grade 40 concrete is $40/3 = 13 \cdot 3$ N/mm².

In this example the three tendons have been placed on the vertical centre line of the section (Figure 5.4) to enable them to be curved upwards through the web. It is necessary to raise a sufficient proportion of the tendons to prevent overstressing of the concrete, either in compression or tension, under conditions of minimum load (i.e. self-weight). The limits within which the resultant prestressing force must lie are defined, as in the previous chapter, by formulae (4.12) to (4.15), illustrated in Figure 5.5, from which formulae (4.12) and (4.14) are seen to be critical. However, when the loading is uniformly distributed, as in this example, and provided the profile of the curved tendons is parabolic, it is sufficient to fix the position of the tendons at the support so that the stress conditions at the support section are satisfied; the line of the prestressing force will then lie completely within the boundaries

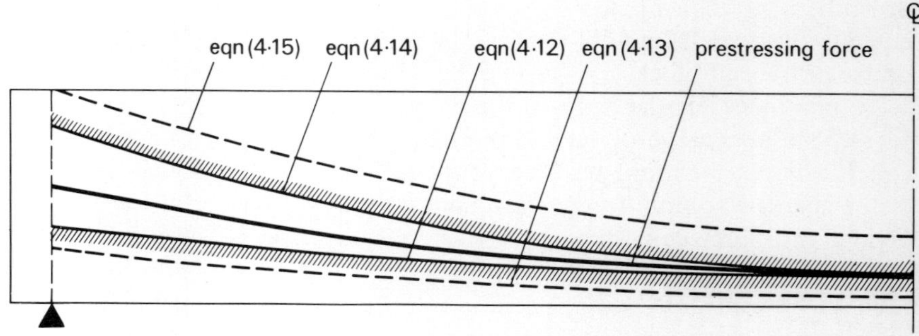

Figure 5.5 Limits of prestressing force

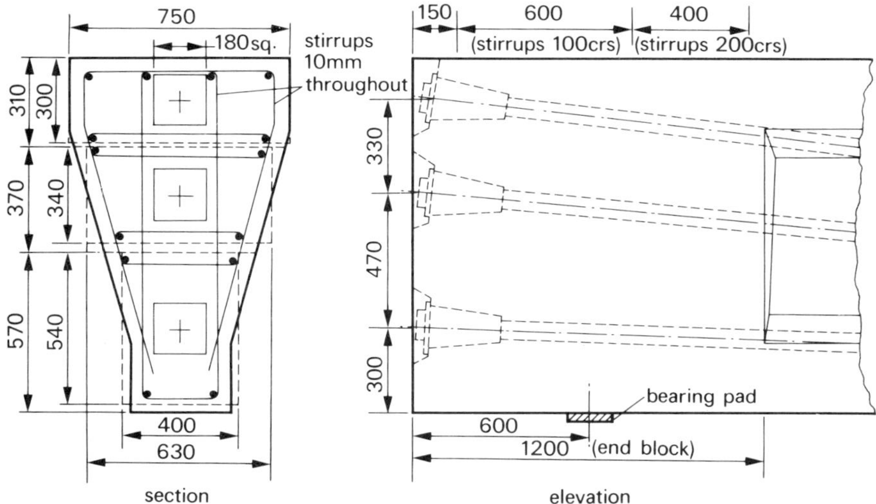

Figure 5.6 Details of end block

representing the stress conditions, since these boundaries are themselves parabolic. The position of the tendons at the supports will be largely determined by a suitable spacing of the anchorages, which will be considered below.

5.4 DESIGN OF END ANCHORAGES

The method of assessing the local stresses and the reinforcement required in the region of the end anchorages has been explained in Chapter 3. In design it is necessary to arrange the anchorages with adequate spacing and cover and in such a way as to permit distribution of the prestressing force over the concrete section in as short a distance and with as little transverse stress as possible. This is facilitated by a solid end block of length approximately equal to the overall depth of the beam. The bearing will normally be near the centre of the end block.

An end block for the beam in the design example is shown in Figure 5.6. Ideally the anchorage of each tendon should be placed so that the force is equal and opposite to the resultant of the stresses on a corresponding portion of the concrete section. In the example the section of the end block has been divided into three equal areas with one anchorage located approximately at the centroid of each area, and the sections of the prisms corresponding to each anchorage force have been drawn so that the force acts at the centre, as shown in Figure 5.6. The bursting force may now be calculated for each prism, as follows:

Bottom anchorage
Vertical bursting force $y_{po}/y_o = 180/540 = 0.33$
$F_{bst} = 0.25 \times 1110 = 278$ kN acting 54–540 mm from end.

Resistance of 5–10 mm stirrups (100 mm centres)
$= 5 \times 2 \times 78 \times 0.87 \times 420 = 285$ kN

Horizontal bursting force $y_{po}/y_o = 180/400 = 0.45$
$F_{bst} = 0.185 \times 1110 = 205$ kN acting 40–400 mm from end.

Resistance of 4–10 mm stirrups (100 mm centres) $= 229$ kN

In the same way the bursting forces are assessed for the centre and top anchorages and it is found that the required resistance can be provided by 10 mm stirrups at 100 mm centres. A possible arrangement of the stirrups is shown in Figure 5.6. Since the bursting forces do not extend over the whole length of the end block the spacing of the stirrups is increased from 100 mm to 200 mm at the end remote from the anchorages.

5.5 CHECK FOR ULTIMATE MOMENT

It is now required to check whether the steel in the tendons provides a sufficient area of tensile reinforcement at the ultimate limit state. The depth of the neutral axis is first calculated from formula (4.24)

$$M_{wu} = M_{ud} - 0.2 f_{cu}(b - b_w)d_f(2d - d_f)$$

$$= 3407 - 0.2 \times 40(750 - 200)300(2 \times 1054 - 300)/10^6$$

$$= 1020 \text{ kNm}$$

$$x/d = 1 - (1 - 5M_{wu}/f_{cu}b_w d^2)^{\frac{1}{2}}$$

$$= 1 - (1 - 5 \times 1020 \times 10^6/40 \times 200 \times 1054^2)^{\frac{1}{2}}$$

$$= 0.347$$

The ultimate resistance of the concrete in compression is obtained from formula (4.25)

$$N_{cu} = 0.4 f_{cu}[(b - b_w)d_f + b_w x]$$

$$= 0.4 \times 40[(750 - 200)300 + 200 \times 0.349 \times 1054]/10^3$$

$$= 3817 \text{ kN}$$

f_{pb}, the design stress in the tendons at the ultimate limit state, is found from CP 110, Table 37 (see Figure 3.6) to be $0.987 \times 0.87 f_{pu}$ when $x/d = 0.347$. Since the tensile strength of one tendon is 1585 kN the total tensile resistance of the tendons is

$$N_{pb} = 3 \times 0.987 \times 0.87 \times 1585 = 4083 \text{ kN}$$

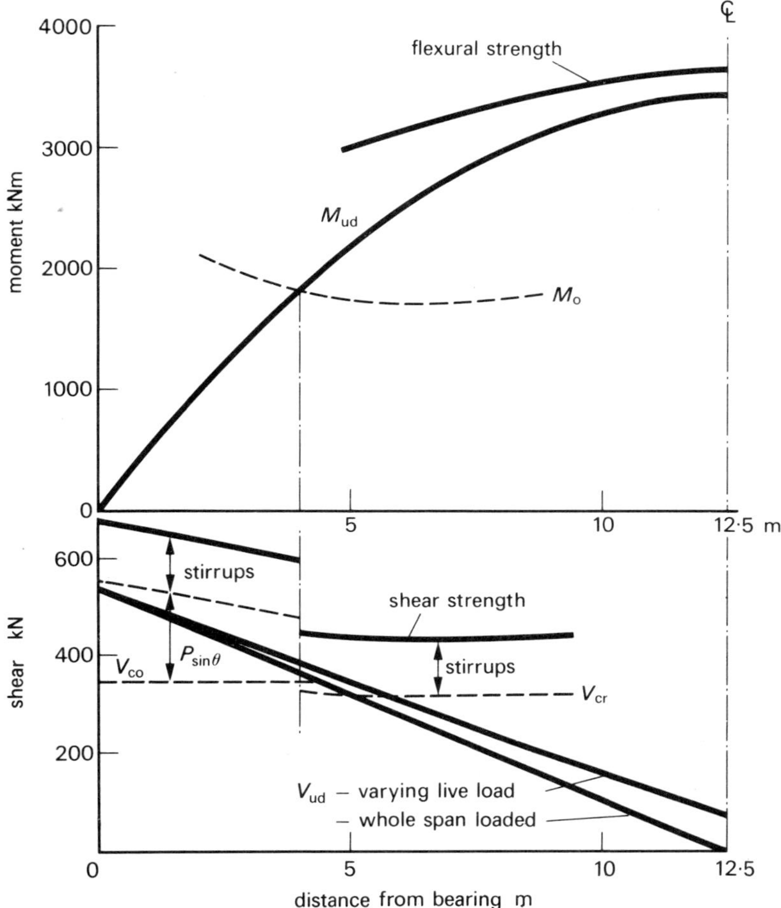

Figure 5.7 Ultimate limit state

This is greater than N_{cu}, and the tensile resistance is therefore adequate. Small diameter reinforcing bars will, however, be provided near the bottom of the beam as a safeguard in the event of accidental cracking due to overloading or to shrinkage before prestressing.

Owing to the parabolic profile of the tendons the ultimate moment of resistance will decrease towards the ends of the beam, but it will always be greater than the design ultimate moment M_{ud} as shown in Figure 5.7. When it is necessary to increase the ultimate moment at mid-span by means of non-prestressed reinforcement, a 'cut-off' diagram may be constructed for these bars as is done for ordinary reinforced concrete.

5.6 CHECK FOR ULTIMATE SHEAR

It is first necessary to find the distance from the support of the beam beyond which the concrete at the level of the tendons will be in tension at the ultimate limit state. Since the level of the tendons varies, the value of the moment M_0 required to cause tension in the concrete must be checked at several points on the beam. In the example, at 2·5 m from the support:

eccentricity of prestressing force

$$e = 231 \text{ mm}$$

effective prestress at level of tendons

$$f_{pt} = \frac{\eta P}{A}\left(1 + \frac{e^2}{i^2}\right)$$

$$= \frac{0\cdot8 \times 3\ 330\ 000}{460\ 000}(1 + 231^2/161\ 500)$$

$$= 7\cdot7 \text{ N/mm}^2$$

$$M_0 = 0\cdot8 f_{pt}\frac{I}{e}$$

$$= 0\cdot8 \times 7\cdot7 \times 74\ 260/231 = 1980 \text{ kNm}$$

The values of M_0 are computed in this way for several points and plotted on the bending moment diagram for the ultimate limit state. This has been done in Figure 5.7, and the intersection of the curves for M_0 and M_{ud} shows the the beam has to be considered to be cracked in flexure at distances greater than 4 m from the supports.

At distances less than 4 m the shear is checked for a region uncracked in flexure. For example, at the support, from CP 110 formula (45) [equation 3.5)]

$$V_{co} = 0\cdot67 b_w h (f_t^2 + 0\cdot8 f_{cp} f_t)^{\frac{1}{2}}$$

tensile strength of concrete $f_t = 0\cdot24(f_{cu})^{\frac{1}{2}}$

$$= 0\cdot24(40)^{\frac{1}{2}} = 1\cdot5 \text{ N/mm}^2$$

compressive stress at centroid $f_{cp} = \dfrac{\eta P}{A}$

$$= 0\cdot8 \times 3330/460\ 000 = 5\cdot8 \text{ N/mm}^2$$

thickness of web (allowing for width of tendon duct),

$$b_{\rm w} = 200 - 64 = 136 \text{ mm}$$

$$V_{\rm co} = 0{\cdot}67 \times 136 \times 1250(1{\cdot}5^2 + 0{\cdot}8 \times 5{\cdot}8 \times 1{\cdot}52/100)^{\frac{1}{2}}$$

$$= 348 \text{ kN}$$

The component of the prestressing force normal to the axis is added to $V_{\rm co}$. At the support the prestressing force is 485 mm above the horizontal tangent at mid-span, and assuming a parabolic profile the slope is

$$2 \times 485/12\,500 = 0{\cdot}0776,$$

Hence $\quad \eta P \sin \theta = 0{\cdot}8 \times 3330 \times 0{\cdot}0776 = 207 \text{ kN}$

Total shear resistance $= V_{\rm co} + \eta P \sin \theta = 348 + 207 = 555 \text{ kN}$

The shear resistance is shown on the ultimate shear diagram in Figure 5.7 and is seen to be adequate without shear reinforcement. It should be remembered that the design ultimate shear $V_{\rm ud}$ is greatest when the live load is placed on one side of the point under consideration.

At distances greater than 4 m from the support the shear resistance is given by formula (46) in CP 110 [equation (3.6)]

$$V_{\rm cr} = \left(1 - 0{\cdot}55\,\frac{f_{\rm pe}}{f_{\rm pu}}\right) v_{\rm c} b_{\rm w} d + M_0\,\frac{V}{M}$$

For example at 5 m from the support:

$$\text{area of tendon } A_{\rm p} = 2910 \text{ mm}^2$$

$$\frac{Ap}{bd} = 2910/200 \times 894 = 1{\cdot}63\,\%$$

From CP 110, Table 5 (Figure 3.7), $v_{\rm c} = 0{\cdot}9 \text{ N/mm}^2$

$$M_0\,\frac{V}{M} = 1711 \times \frac{4}{25} = 274 \text{ kN}$$

$$V_{\rm cr} = (1 - 0{\cdot}55 \times 0{\cdot}8 \times 3330/4750)0{\cdot}88 \times 136 \times 894/1000 + 274$$

$$= 348 \text{ kN}$$

The values of $V_{\rm cr}$ are similarly calculated for other points on the span and are shown in Figure 5.7. $V_{\rm cr}$ is found to be less than the ultimate shear between about 4 m and 4·5 m from the supports, and reinforcement must be provided over this length.

Area of shear reinforcement required per unit length [CP 110, formula (48)] [equation (3.8)],

$$\frac{A_{sv}}{s_v} = \frac{V_{ud}V_{cr}}{0.87f_{yv}d_t}$$

$$= (380 - 348)1000/(0.87 \times 410 \times 1220) = 0.074 \text{ mm}^2/\text{mm}$$

This, however, is less than the minimum shear reinforcement given by CP 110, formula (47) [equation (3.7)]

$$\frac{A_{sv}}{s_v} = \frac{0.4b_w}{0.87f_{yv}}$$

$$= 0.4 \times 200/(0.87 \times 410) = 0.224 \text{ mm}^2/\text{mm}$$

The maximum spacing of the stirrups is $0.75d_t$ (where d_t is the depth to the bottom bar to which the stirrups are anchored) or $4b_w$. Allowing 25 mm cover, $d_t = 1220$ mm, so that the spacing must not exceed $0.75 \times 1220 = 915$ mm or $4 \times 200 = 800$ mm.

Using 10 mm stirrups at 600 mm centres,

$$\frac{A_{sv}}{s_v} = 157/670 = 0.262 \text{ mm}^2/\text{mm}$$

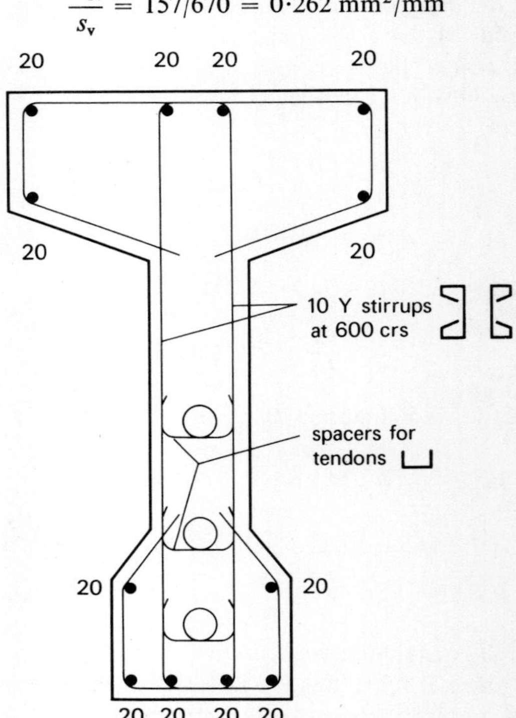

Figure 5.8 Shear reinforcement

The resistance of these stirrups is added to V_{cr}, or to $V_{co}+P\sin\theta$, to give the total shear strength as in Figure 5.7. According to CP 110 stirrups are not required where V_{ud} is less than $0\cdot5V_c$, but in a beam of this type they are better retained as a safeguard in the event of shrinkage cracking, and to support the tendons.

A possible arrangement of stirrups is shown in Figure 5.8. Each stirrup consists of two parts wired together and forming a cage with 20 mm longitudinal bars. Spacers may be wired to the legs of the stirrups to support the sheathed tendons, or the tubes forming the ducts; alternatively these may be supported on spacing bars passing through the sides of the mould, which are withdrawn after placing the concrete.

6

Composite Beams of Precast and Cast In-Situ Concrete

6.1 COMPOSITE CONSTRUCTION

Composite construction is an economical method of increasing the cross-sectional area and hence stiffening and strengthening a prestressed beam by the addition of plain concrete. Concrete may also be added to form a continuous structure from a series of prestressed beams as discussed in the following chapter. In most practical applications the original prestressed beam is precast away from the site and the additional concrete placed *in situ*.

One of the commonest types of composite construction is that in which a number of prestressed precast beams, placed side by side, are connected by a continuous top slab placed *in situ*. In floor construction this technique is a development of the well-known practice of using reinforced concrete ribs, either precast or cast *in situ*, connected by a topping of *in situ* concrete which may be supported on hollow blocks. Rules governing the minimum thickness of the topping are given in CP 110, 3.7.12.

On a large scale, a composite bridge deck may be formed by means of I section prestressed beams connected by the top slab, placed *in situ*, an example of which may be seen in Figure 6.3. This type of deck has little stiffness in the transverse direction and in order to provide for the effective distribution of concentrated loads the prestressed beams are often connected by a series of transverse beams, ribs or diaphragms. These may be formed by means of stubs cast as part of the main beams, and connected by prestressing, or they may consist of reinforced or prestressed concrete beams cast *in situ*. Whatever the method of construction, however, transverse beams are a source of delay and expense, and the current trend is to increase the transverse stiffness by providing a thicker top slab with appropriate reinforcement.

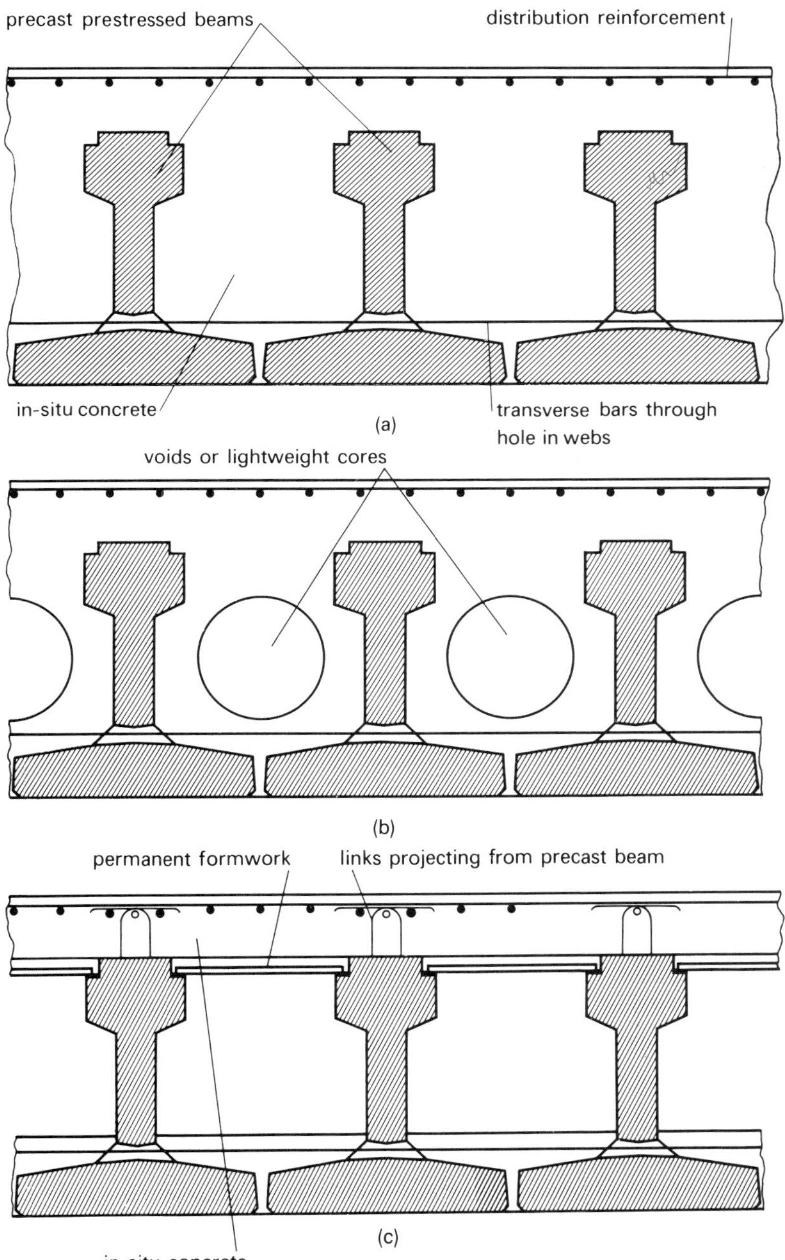

Figure 6.1 Composite bridge decks with precast prestressed inverted T beams

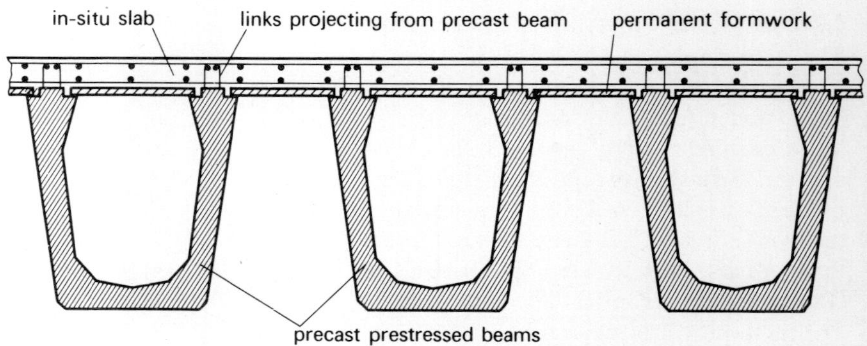

Figure 6.2 Composite bridge deck with precast prestressed U beam

The fresh *in situ* concrete may be supported by conventional formwork or by permanent formwork of asbestos-cement or fibre-reinforced concrete. However, since formwork constitutes an important element in the cost of concrete construction it is a great advantage if the precast beam itself can be used as permanent formwork. This is done in the various types of deck incorporating prestressed beams of inverted T section, examples of which are shown in Figure 6.1. The solid slab Figure 6.1(a) is simple to construct but a lighter hollow deck may be cast using tubular void formers or cores of a light-weight material such as expanded polystyrene [Figure 6.1(b)]. Another method is to connect the bottom flanges of the inverted T beams with a layer of *in situ* concrete and to cast a separate top slab on permanent formwork as in Figure 6.1(c), thus forming what is sometimes termed a pseudo-box construction.

An alternative technique to minimize the bending stress due to the dead weight of the slab is to prop the prestressed beams while placing and curing the *in situ* concrete. This will be explained in detail later.

The analysis of longitudinal and transverse moments induced in composite slab bridge decks by concentrated loads, such as the Ministry of Transport HB loading, is often accomplished by the analogy of an orthotropic plate and the method and its application have been described in several publications, refs. (6.1–6.4). It is not, however, suitable for bridge decks with a skew greater than 20°. In such cases a grillage analysis, ref. (6.5), or a finite element computer programme, ref. (6.6) may be used.

Prestressed beams of U section are sometimes used at the edges of a composite deck, but it has recently been claimed that the most economical deck consists of U beams connected by a top slab, ref. (6.7) (Figure 6.2). This design combines, to some extent, the lightness of the beam and slab deck with the torsional stiffness of the multi-cellular box deck.

6.2 CALCULATION OF STRESSES IN CONCRETE

The dimensional properties of the precast prestressed beam must obviously be used for the calculation of all stresses induced in the beam before the *in situ* concrete has hardened; namely, stresses due to prestressing and bending stresses induced by the weight of the beam itself, the fresh *in situ* concrete and the formwork except when the latter is supported independently of the prestressed beam. However, when the *in situ* concrete has hardened a larger composite section has been formed, and stresses resulting from subsequent loading must be calculated on the basis of the dimensional properties of the composite section and added to the existing stresses in the prestressed beam. The method will be clear from an example.

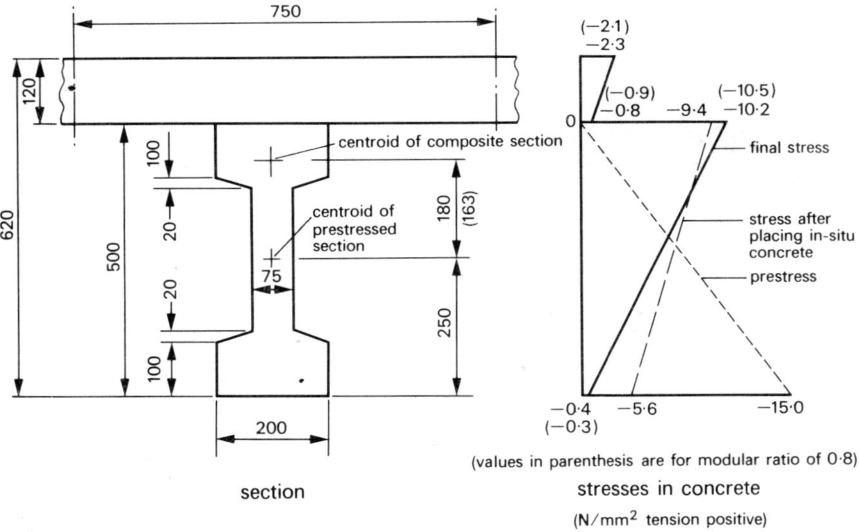

Figure 6.3 Stresses in a composite section

Example

A composite bridge deck, the cross-section of which is shown in Figure 6.3, has a span of 12 m *and is formed from prestressed symmetrical I beams at* 750 mm *centres with a top slab of* in situ *concrete* 120 mm *thick. If the compressive prestress in the beams is* 15 N/mm² *at the bottom and zero at the top, calculate the stresses in the concrete under an imposed load of* 5 kN/m².

The dimensional properties of the prestressed precast section are as follows

$$\text{area } A_1 = 65\,000 \text{ mm}^2$$

$$\text{second moment of area } I_1 = 1849 \times 10^6 \text{ mm}^4$$

$$\text{section modulus (top and bottom) } Z_1 = 1849 \times 10^6/250$$
$$= 7\cdot40 \times 10^6 \text{ mm}^3$$

The load supported by the prestressed beam alone is as follows:

$$\begin{aligned}
\text{self-weight: } 0{\cdot}065 \times 2400 \times 9{\cdot}81/1000 &= \quad 1{\cdot}53 \text{ kN/m}\\
\textit{in situ} \text{ concrete: } 0{\cdot}75 \times 0{\cdot}12 \times 23{\cdot}5 &= \quad 2{\cdot}12 \text{ kN/m}\\
\text{allow for formwork:} &= \quad 0{\cdot}20 \text{ kN/m}\\
\hline
&= \quad 3{\cdot}85 \text{ kN/m}
\end{aligned}$$

$$\text{moment } M_1 = 3{\cdot}85 \times 12^2 \times 0{\cdot}125 \quad = 69{\cdot}3 \text{ kN/m}$$

The compressive stresses in the concrete at the bottom and top of the prestressed beam after placing the *in situ* concrete are

$$f_{\text{inf}} = 15 - 69{\cdot}3/7{\cdot}40 = 5{\cdot}6 \text{ N/mm}^2$$
$$f_{\text{sup}} = 0 + 69{\cdot}3/7{\cdot}40 = 9{\cdot}4 \text{ N/mm}^2$$

For the next stage it is necessary to calculate the dimensional properties of the composite section comprising the beam and the corresponding 750 mm width of the top slab. If the *in situ* concrete has the same modulus of elasticity as the beam, these are

area $A_2 = 65\,000 + 750 \times 120 = 155\,000 \text{ mm}^2$
height of centroid above centroid of prestressed beam
$\quad = 90\,000(250 + 120/2)155\,000 = 180 \text{ mm}$
second moment of area,
$\quad I_2 = 1849 \times 10^6 + 65\,000 \times 180^2 + 90\,00[120^2/12 + (250 + 120/2 - 180)^2]$
$\quad\quad = 5584 \times 10^6 \text{ mm}^4$

Section moduli for prestressed part of composite section,
$\quad Z_{2\text{p inf}} = 5584 \times 10^6/(250 + 180) = 13{\cdot}0 \times 10^6 \text{ mm}^3$
$\quad Z_{2\text{p sup}} = 5584 \times 10^6/(250 - 180) = 79{\cdot}8 \times 10^6 \text{ mm}^3$

Section moduli for *in situ* concrete in composite section,
$\quad Z_{2\text{i inf}}$ (as for top of prestressed section) $= 79{\cdot}8 \times 10^6 \text{ mm}^3$
$\quad Z_{2\text{i sup}} = 5584 \times 10^6/(250 + 120 - 180) = 29{\cdot}4 \times 10^6 \text{ mm}^3$
load supported by composite section
$\quad = 5 \times 0{\cdot}75 = 3{\cdot}75 \text{ kN/m}$
moment $M_2 = 3{\cdot}75 \times 12^2 \times 0{\cdot}125 = 67{\cdot}5 \text{ kNm}$

The final stresses in the composite section may now be calculated. Prestressed section

$$f_{\text{inf}} = 5{\cdot}6 - 67{\cdot}5/13{\cdot}0 = 0{\cdot}4 \text{ N/mm}^2 \text{ (compression)}$$
$$f_{\text{sup}} = 9{\cdot}4 + 67{\cdot}5/79{\cdot}8 = 10{\cdot}2 \text{ N/mm}^2 \text{ (compression)}$$

In situ concrete

$$f_{inf} = 67\cdot5/79\cdot8 = 0\cdot8 \text{ N/mm}^2 \text{ (compression)}$$
$$f_{sup} = 67\cdot5/29\cdot4 = 2\cdot3 \text{ N/mm}^2 \text{ (compression)}$$

The stress distribution diagrams are shown in Figure 6.3. It will be noted that the severest stresses occur in the prestressed part of the section and that the *in situ* concrete in the top slab is subjected only to low compressive stresses. For this reason a lower grade of concrete is often specified for the *in situ* concrete in a composite section; in any event the *in situ* concrete will be younger and likely to have a somewhat lower modulus of elasticity than the precast prestressed concrete. The modular ratio of the concrete in the two parts is not likely to be less than about 0·8, and the effect can be taken into account in the calculation of the dimensional properties of the composite section by reducing the area of the *in situ* concrete in the ratio of its modulus of elasticity to that of the prestressed concrete. The section thus obtained represents an equivalent area of precast prestressed concrete.

In the previous example, if the modular ratio were 0·8, the calculation would be as follows

area $A_2 = 65\,000 + 0\cdot8 \times 750 \times 120 = 137\,000 \text{ mm}^2$
height of centroid above centroid of prestressed beam
$$= 72\,000(250 + 120/2)/137\,000 = 163 \text{ mm}$$
second moment of area,
$$I_2 = 1849 \times 10^6 + 65\,000 \times 163^2 + 72\,000[120^2/12 + (250 + 120/2 - 163)^2]$$
$$= 5218 \times 10^6 \text{ mm}^4$$

Section moduli for prestressed part of composite section,
$$Z_{2p\,inf} = 5218 \times 10^6/(250 + 163) = 12\cdot6 \times 10^6 \text{ mm}^3$$
$$Z_{2p\,sup} = 5218 \times 10^6/(250 - 163) = 60\cdot0 \times 10^6 \text{ mm}^3$$

Section moduli for *in situ* concrete in composite section,
$$Z_{2i\,inf} = 5218 \times 10^6/0\cdot8(250 - 163) = 75\cdot0 \times 10^6 \text{ mm}^3$$
$$Z_{2i\,sup} = 5218 \times 10^6/0\cdot8(250 + 120 - 163) = 31\cdot5 \times 10^6 \text{ mm}^3$$

It will be noted that the latter section moduli have been related to the area of the *in situ* concrete by dividing by the modular ratio, so as to give the stresses in the *in situ* concrete.

The final stresses in the composite section are calculated as before and are slightly modified on account of the change in the section moduli. The values obtained are given in parenthesis in Figure 6.3.

If the prestressed beam is propped while placing the *in situ* concrete, or if the latter is supported in some other way independent of the prestressed beams, the load due to the *in situ* concrete and permanent formwork is later supported by the composite member instead of being supported during construction by the prestressed beam alone. The moment acting on the latter will then be,

$$M_1 = 1 \cdot 53 \times 12^2 \times 0 \cdot 125 = 27 \cdot 5 \text{ kNm}$$

Moment supported by composite section,

$$M_2 = (3 \cdot 75 + 2 \cdot 12 + 0 \cdot 20) \times 12^2 \times 0 \cdot 125 = 109 \cdot 3 \text{ kNm}$$

Under these conditions the severity of the total bending stresses will be somewhat reduced. For example, the final stress at the soffit of the prestressed beam will be given by

$$f_{\text{inf}} = 15 - 27 \cdot 5/7 \cdot 40 - 109 \cdot 3/12 \cdot 6 = 2 \cdot 61 \text{ N/mm}^2 \text{ (compression)}$$

6.3 SERVICEABILITY LIMIT STATE—ALLOWABLE STRESSES IN CONCRETE

The allowable compressive and tensile stresses in the prestressed and *in situ* concrete of a composite section at the serviceability limit state are related to the strength of the concrete in the respective parts. Generally these stresses are based on the normal rules for prestressed concrete (e.g. CP 110, 4.3.3.2 and 4.3.3.3 and ACI 318–71, 18.4) which have been discussed in Chapter 2, but exceptions may be made in respect of high stresses in two situations.

The first of these concerns the compressive stress in the precast prestressed unit at the interface with the *in situ* concrete. The value of this stress is sometimes high, particularly at the top of the inverted T type of precast section as will be seen from the example in Figure 6.4. According to CP 110, 5.4.3.2, the normal allowable compressive stress may be increased above the values given in Table 32 by not more than 50% provided the ultimate failure of the composite beam would be due to excessive elongation of the steel. The latter requirement is a safeguard against the undesirable compressive failure of the concrete at the ultimate limit state. The only objection to a high compressive stress at the serviceability limit state is the possibility of a large deformation due to creep; this would not apply in this instance since the highly stressed concrete forms a relatively small part of the total zone of compression.

The other exception to the normal allowable stress conditions is contained in CP 110, 5.4.3.3 (Table 2.6) and concerns high flexural tensile stresses in the *in situ* concrete at the contact surface with the prestressed concrete. These stresses occur in composite slabs with inverted T beams, as in Figure 6.4. The allowable tensile stresses, which are given in Table 44 (Table 2.6) range from 3·2 N/mm² to 5·0 N/mm² according to the grade of the *in situ* concrete and are considerably greater than the corresponding allowable stresses for prestressed concrete members since it has been proved experimentally that the development of visible cracking is inhibited by the uncracked prestressed concrete to which the *in situ* concrete is bonded.

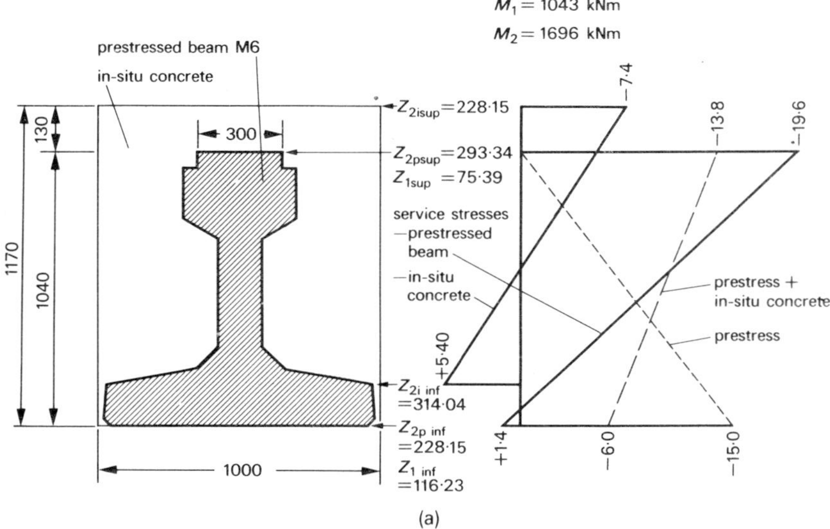

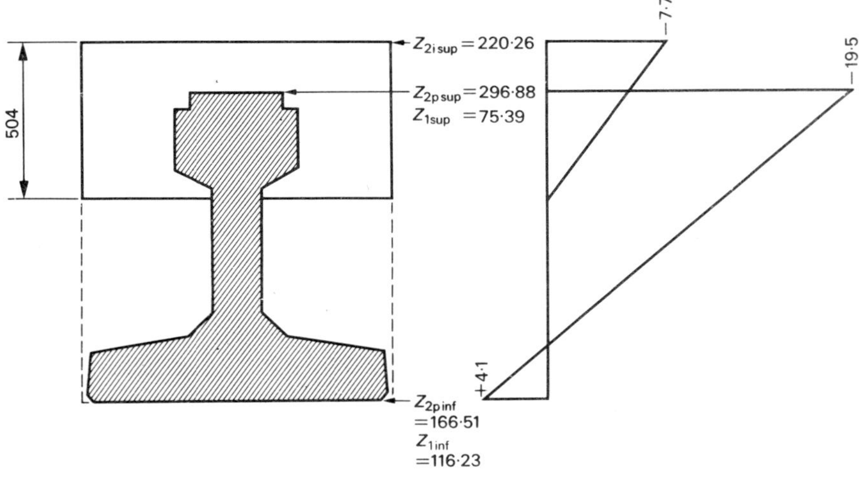

Figure 6.4 Stresses in uncracked and cracked composite section: (a) uncracked section; (b) cracked section

The above allowable stresses may be further increased by 50% provided the permissible tensile stress in the precast concrete unit is reduced by the same numerical amount. The reasoning underlying this provision is that the existence of cracks in the *in situ* concrete would, by reducing the stiffness of the section, lead to an increased tensile stress at the soffit of the prestressed beam and hence to earlier cracking unless the allowable stress, calculated assuming the *in situ* concrete to be uncracked, is reduced.

An example with calculated stresses is shown in Figure 6.4. A C and CA standard bridge beam M6 is incorporated in a solid composite slab 1170 mm deep. If the prestressed concrete is of Grade 50 and the *in situ* concrete of Grade 30, the allowable tensile stresses will be $3 \cdot 2$ N/mm^2 and $3 \cdot 6$ N/mm^2 respectively. The tensile stress in the *in situ* concrete at the interface may be increased by 50% to $5 \cdot 4$ N/mm^2, i.e. by $1 \cdot 8$ N/mm^2, if the tensile stress at the soffit of the prestressed beam is reduced to $3 \cdot 2 - 1 \cdot 8 = 1 \cdot 4$ N/mm^2. Typical stress distribution diagrams are shown in Figure 6.4(a), assuming the prestress in the beam to be $15 \cdot 00$ N/mm^2 at the bottom and zero at the top.

Figure 6.4(b) shows the corresponding stress diagram calculated for the same moments acting on the prestressed beam alone and on the cracked composite section. Assuming the *in situ* concrete to resist no tensile stress the depth of the neutral axis is found to be 504 mm and the tensile stress at the soffit of the prestressed beam is $4 \cdot 1$ N/mm^2, somewhat greater than the original allowable stress of $3 \cdot 2$ N/mm^2.

6.4 ULTIMATE LIMIT STATE—FLEXURE

In checking a composite beam at the ultimate limit state the ultimate moment may be calculated according to the same principles as an ordinary prestressed beam. The compression zone will often consist entirely of *in situ* concrete, and the value of f_{cu} to be used will obviously be that of the *in situ* concrete. Sometimes, however, the compression zone will contain part of the precast beam, in which event the calculation may be adjusted for greater accuracy.

For example, in the composite slab in Figure 6.4, if all the load applied after the *in situ* concrete has hardened is considered to be live load, the design value of the ultimate moment will be

$$M_{ud} = 1 \cdot 4 \times 1043 + 1 \cdot 6 \times 1696 = 4174 \text{ kNm}$$

Assuming, initially, that the whole of the compression zone lies in the *in situ* concrete and that the tensile reinforcement is located at the centre of the bottom flange, the depth x of the compression zone is calculated from formula (4.21).

$$M_{ud}/f_{cu}bd^2 = 4174 \times 10^6/(30 \times 1000 \times 1090^2) = 0 \cdot 117$$

$$x/d = 1 - [1 - 5(M_{ud}/f_{cu}bd^2)]^{\frac{1}{2}} = 1 - (1 - 5 \times 0 \cdot 117)^{\frac{1}{2}} = 0 \cdot 356$$

$$x = 0 \cdot 356 \times 1090 = 388 \text{ mm}$$

It will be seen from the diagram that the top flange of the precast pre-stressed beam occupies about one-fifth of the compression zone; a suitable average strength of the concrete would therefore be

$$50 \times 1/5 + 30 \times 4/5 = 34 \text{ N/mm}^2$$

Recalculating with

$$f_{cu} = 34 \text{ N/mm}^2, \quad x = 331 \text{ mm}$$

The moment of resistance may be checked:

$$M_u = 0.4 \times 30 \times 1000 \times 331(1090 - 331/2)$$
$$+ 0.4(50 - 30)400 \times 201(1090 - 130 - 201/2) \text{ Nmm}$$
$$= 4225 \text{ kNm} \ (\simeq 4174 \text{ kNm})$$

The compressive resistance of the concrete is

$$N_{cu} = 0.4 \times 30 \times 1000 \times 331 + 0.4(50 - 30)400 \times 201 \text{ N}$$
$$= 4615 \text{ kN}$$

This force must be balanced by the tensile resistance of the tendons and additional reinforcement (if required) at the ultimate limit state.

6.5 ULTIMATE LIMIT STATE—SHEAR

To check the shear strength of composite sections uncracked in flexure the shear stress is calculated in the same two stages that were considered when calculating the bending stresses, namely for the load acting on the precast

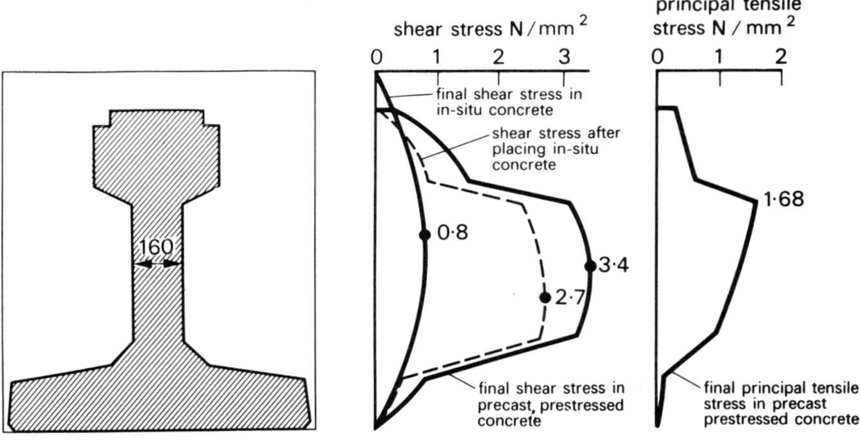

Figure 6.5 Shear and principal tensile stresses in a composite section

prestressed section alone, and for the additional load acting on the composite section. Here, however, the loads must be enhanced by the factors specified for the ultimate limit state. The total shear stress is then used in conjunction with the stress normal to the section (prestress + bending stress) to compute the principal tensile stress, which must not exceed the design tensile strength of the concrete.

Using, again, the example in Figure 6.4, if the span of the beam is 17·5 m the characteristic shear forces at the support will be 238 kN acting on the precast prestressed member alone, and a further 388 kN live load acting on the composite section. At the ultimate limit state the respective shear forces will therefore be $V_1 = 1·4 \times 238 = 333$ kN and $V_2 = 1·6 \times 388 = 621$ kN.

The distribution of shear stress in the precast section due to a shear force of 333 kN may be calculated by the formula $v_1 = VS/Ib$ and is shown in Figure 6.5. The additional shear stress caused by a shear force of 621 kN acting on the composite section may similarly be calculated; since the composite section is rectangular if the modular ratio is unity, the shear stress distribution will be parabolic with a maximum value

$$v_2 = \frac{1·5 \times 621\ 000}{1000 \times 1170} = 0·80\ \text{N/mm}^2.$$

The latter shear distribution applies to the *in situ* concrete, and is added to the former to give the total shear stress in the precast prestressed section, as in Figure 6.5. Finally the principal tensile stress f_{max} in the precast prestressed section is calculated from the expression:

$$f_{max} = -\frac{f_x}{2} + \left[\frac{f_x}{2} + (v_1 + v_2)^2 \right]^{\frac{1}{2}}$$

$$= \frac{f_x}{2} \left\{ -1 + \left[1 + 4(v_1 + v_2)^2/f_x^2 \right]^{\frac{1}{2}} \right\}$$

f_x is the compressive stress normal to the section. At the support there is no bending stress and f_x is therefore equal to the prestress. The distribution of the principal tensile stress is shown in Figure 6.5. In the *in situ* concrete there is no normal stress and the principal tensile stress is numerically equal to the shear stress. It will be seen from the stress distribution diagrams that the maximum values of the shear stress in the two stages and the principal tensile stress all occur at different points. A simplified approximate calculation may be made by the method adopted in CP 110, 4.3.5.1, calculating the principal tensile stress at the centroid of the composite section, assuming the shear stress to be $V/0·67bh$ in each of the two stages, where b is the breadth at the centroid and h the total depth of the appropriate section. The calculation would then be as follows:

shear stress

prestressed section alone $v_1 = 333 \times 10^3/(0.67 \times 160 \times 1040) = 3.0 \text{ N/mm}^2$
composite section $\qquad v_2 = 621 \times 10^3/(0.67 \times 1000 \times 1170) = 0.8 \text{ N/mm}^2$

total $\qquad\qquad v_1 + v_2 = \qquad\qquad\qquad\qquad\qquad 3.8 \text{ N/mm}^2$

normal stress = prestress at centroid of composite section

$$f_x = 15.00(1040 - 1170/2)1040 = 6.6 \text{ N/mm}^2$$

principal tensile stress,

$$f_{\max} = (6.6/2)\{-1 + [1 + 4(3.8/6.6)^2]^{\frac{1}{2}}\} = 1.7 \text{ N/mm}^2$$

According to CP 110, 4.3.5.1 the tensile strength of the concrete,

$$f_t = 0.24(f_{cu})^{\frac{1}{2}} = 0.24(50)^{\frac{1}{2}} = 1.7 \text{ N/mm}^2.$$

This is approximately equal to the principal tensile stress calculated above; if, however, the normal stress is reduced to 0.8 times the prestress at the centroid as recommended in CP 110 the value of the principal tensile stress becomes 1.98 N/mm^2 and web reinforcement must be provided. To calculate the amount required it is first necessary to determine V_{co}, the total shear at which the principal tensile stress is 1.7 N/mm^2.

$$v_{co} = v_1 + v_{2c} = (f_t^2 + 0.8 f_{cp} f_t)^{\frac{1}{2}} = 0.67(1.7^2 + 0.8 \times 6.6 \times 1.7)^{\frac{1}{2}}$$

$$= 3.4 \text{ N/mm}^2$$

$$v_{2c} = 3.4 - 3.0 = 0.4 \text{ N/mm}^2$$

$$V_{2c} = 0.67 \, bh v_{2c} = 0.67 \times 1000 \times 1170 \times 0.43/1000$$

$$= 337 \text{ kN}$$

$$V_{co} = V_1 + V_{2c} = 333 + 337 = 670 \text{ kN}$$

Then from CP 110, 4.3.5.3, formula (48):

$$\frac{A_{sv}}{s_v} = \frac{V - V_{co}}{0.87 f_{yv} d_t} = \frac{[(333 + 621) - 670] \times 1000}{0.87 \times 410 \times 0.9 \times 1170} = 0.756 \text{ mm}^2/\text{mm}$$

(using steel of yield stress 410 N/mm^2 and assuming $d_t = 0.9h$).

This may be provided by 10 mm stirrups at 200 mm centres ($0.785 \text{ mm}^2/\text{mm}$).

It should be noted that from CP 110, 4.3.5.3, formula (47), the minimum shear reinforcement for this beam is,

$$\frac{A_{sv}}{s_v} = \frac{0.4 b_w}{0.87 f_{yv}} = \frac{0.4 \times 160}{0.87 \times 410} = 0.179 \text{ mm}^2/\text{mm}$$

This is based on the breadth of the web of the precast prestressed beam which would appear to be a reasonable interpretation of the Code since it is in the prestressed web that inclined cracking is likely to occur and the stirrups are normally located.

The shear strength of composite beams in regions which would be cracked in flexure under the design ultimate moment may be calculated on the basis of CP 110, 4.3.5.2 or ACI 318–71 11.5.2. M_0, the moment necessary to produce zero stress in the concrete at the level of the tendons will be given by

$$M_0 = M_1 \left(1 - \frac{y_1 I_2}{y_2 I_1} \right) + \frac{0 \cdot 8 f_{pt} I_2}{y_2}$$

where M_1 = moment acting on prestressed section alone,
　　　I_1 = second moment of area of prestressed section,
　　　y_1 = distance of centroid of tendons from centroid of prestressed section,
　　　I_2 = second moment of area of composite section,
　　　y_2 = distance of centroid of tendons from centroid of composite section,
　　　f_{pt} = prestress in concrete at level of tendons.

Formula (46) [equation (3.6)] may then be used to calculate V_{cr} taking the values of b (breadth) and d (distance from compression face to centroid of tendons) from the composite section. The ultimate shear stress v_c, obtained from Table 5 in CP 110, is also based on the composite section.

If the design ultimate shear at the section under consideration exceeds V_{cr}, shear reinforcement is designed for the composite section by the normal method.

6.6 SHEAR AT THE INTERFACE

The composite action upon which the stress calculations have been based is dependent on an effective shear connection across the surface of contact between the precast and the *in situ* concrete. It is assumed that the natural bond at the interface can develop a certain level of shear resistance, according to the strength of the *in situ* concrete and the roughness of the precast surface; beyond this, steel shear connectors will be required.

In CP 110, 5.4.3.4 it is recommended that the shear connection should be designed for the service load (characteristic load) condition for the total shear. This is questionable since the strength of the shear connection must be maintained up to the ultimate load, suggesting that an ultimate limit state calculation would as in ACI 318–71, 17.5.3 be more logical. In addition, shear stress can only be developed at the interface after hardening of the *in situ* concrete, so that in calculating the shear the component of the dead load G_k due to the weight of the prestressed beam and *in situ* concrete should not be included except when the beam has been propped during construction.

The horizontal shear stress v_h due to a design shear V_d is given by the standard shear stress formula (see Chapter 3, pp. 39–40).

$$v_h = \frac{V_d S_c}{I b_e}$$

where S_c = first moment of area of the concrete to one side of the composite section about the centroidal axis of the transformed composite section

I = second moment of area of composite section

b_e = width of contact surface.

The shear stresses allowed by CP 110 are set out in 5.4.3.4, Table 45 and in ACI 318–71, 17.5.4. In CP 110 the stresses are slightly more restrictive for beam and slab construction (Figure 6.3) than for solid composite slabs [Figure 6.1(a)]. The stresses permitted without connecting links are seen from Figure 3.9 to be greatest for 'Type 1' surfaces in which the precast surface is roughened by brushing or by a water spray before the concrete has hardened. A smooth untreated contact surface is only permitted in beam and slab construction when links are provided (see pp. 39–40).

Figure 3.9 also shows the allowable stresses in composite construction with links. The minimum reinforcement is 0.15% of the contact surface, and this must be provided if the surface of the precast beam is smooth and untreated (Type 2). Higher stresses are permitted with a roughened (Type 3) surface, and these may be further increased by $0.5\ \text{N/mm}^2$ for each additional one per cent of links.

Example

(1) *Check the shear connection at the interface of the composite beam and slab section Figure 6.3, at the support.*

V_d (see p. 89–90) $= 3.75 \times 12 \times 0.5 = 22.5\ \text{kN}$

$$S_c = 750 \times 120(620 - 250 - 180 - 120/2) = 11.70 \times 10^6\ \text{mm}^3$$

$$b_e = 200\ \text{mm}$$

$$v_h = \frac{V_d S_c}{I b_e} = \frac{22.5 \times 10^3 \times 11.70 \times 10^6}{5584 \times 10^6 \times 200} = 0.24\ \text{N/mm}^2$$

In this and the following calculation the modular ratio of the two concretes has been assumed to be 1.0. If it is reduced to 0.8 the shear stress will be slightly less severe.

From CP 110 5.4.3.4, or from Figure 3.9 it is seen that for all grades of *in situ* concrete, the required shear resistance may be provided by a Type 1 roughened surface or by a Type 2 untreated surface with the minimum area of links. The latter may be provided by 6 mm links at 180 mm centres, the area of which is 0·157% of the contact area.

(2) *Check the shear connection at the interface of the composite slab section shown in Figure 6.4(a) at the support.*

V_d (see p. 96) $= 388$ kN.

The critical face for contact shear will be the top of the precast beam.

$$S_c = 1000 \times 130(1170/2 - 130) = 59 \cdot 15 \times 10^6 \text{ mm}^3$$

$$v_h = \frac{V_d S_c}{I b_e} = \frac{388 \times 10^3 \times 59 \cdot 15 \times 10^6}{133\,468 \times 10^6 \times 300} = 0 \cdot 57 \text{ N/mm}^2$$

From CP 110, 5.4.3.4 or Figure 3.9 the shear resistance for Grade 30 concrete without links is found to be 0·54 N/mm² for a Type 1 surface or 0·30 N/mm² for a Type 2 surface. With the minimum area of links, however, these values are increased to 1·25 N/mm² for a roughened (Type 2) surface or 0·38 N/mm² for Type 2. The first of these two is therefore adopted.

The shear reinforcement provided in the precast prestressed beam should normally extend into the *in situ* concrete and may be considered to act as connecting links. The shear reinforcement in this beam (p. 97) consists of 10 mm stirrups at 200 mm centres; the area is $78 \cdot 5 \times 2/(300 \times 200) = 0 \cdot 26\%$ of the contact area and is therefore sufficient for the shear connection.

6.7 DIFFERENTIAL SHRINKAGE

Shrinkage, unless retarded by moist curing, occurs most rapidly in the early life of concrete and consequently a considerable proportion of the total shrinkage of a precast prestressed beam may occur before a composite section is formed by the hardening of additional *in situ* concrete. On the other hand the *in situ* concrete forms part of the composite section for the whole of its life and may therefore be expected to undergo a greater effective shrinkage which may be further increased if the water content is appreciably greater than that of the precast concrete. The differential shrinkage of the two parts of the composite section will induce stresses in both the precast and the *in situ* concrete.

The shrinkage of concrete varies greatly according to its composition and environment and in critical cases it may be necessary to obtain data from tests. In the absence of more exact information, however, a general value of 100×10^{-6} is suggested in CP 110, 5.4.3.5 for differential shrinkage calculations.

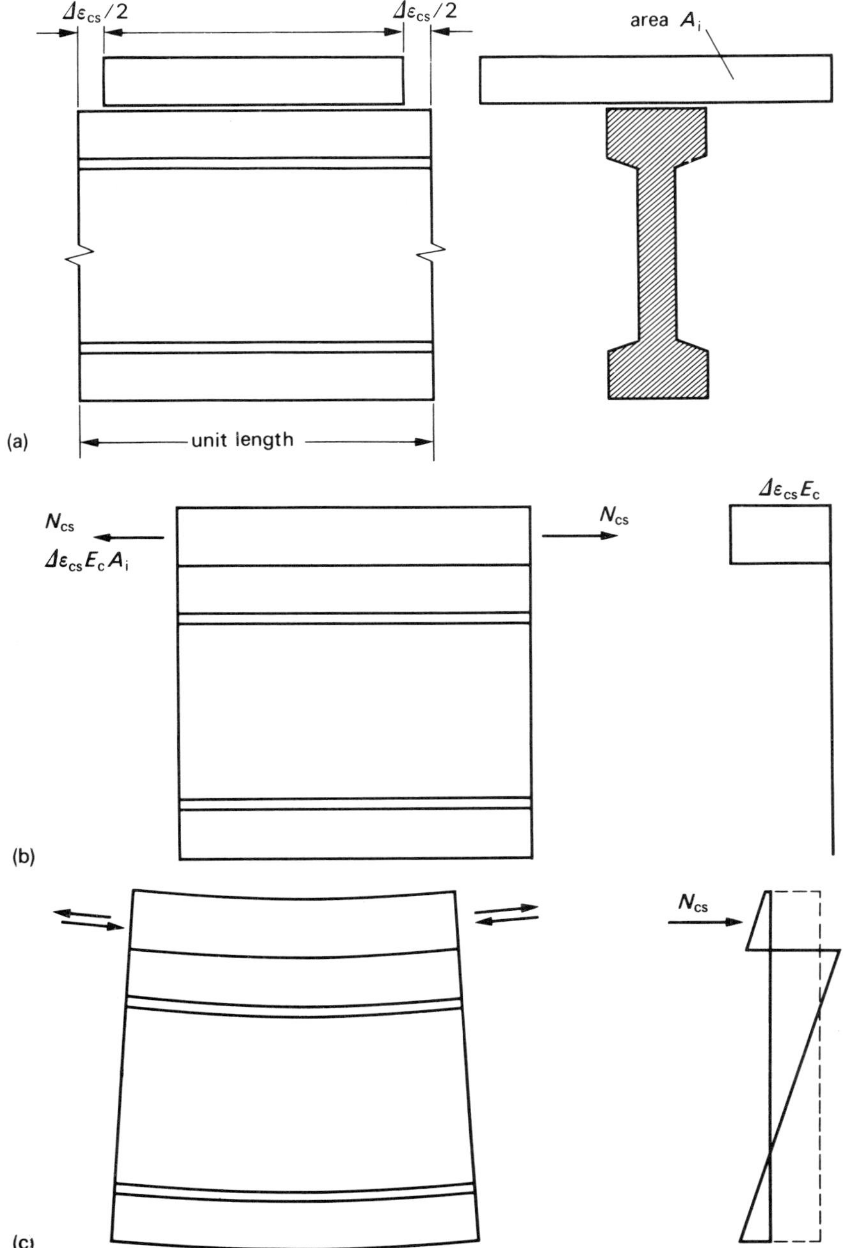

Figure 6.6 Calculation of stresses due to differential shrinkage

The following method of calculating the stresses due to differential shrinkage assumes that it is uniform over the *in situ* part of the section and is unaffected by other time-dependent effects, such as, for example, creep, increase in modulus of elasticity with age and the component of shrinkage which is common to the two parts of the section. It may best be illustrated by the example of a beam and slab section.

If the top slab were considered not to be connected to the precast element it would be free to undergo the full amount of differential shrinkage $\Delta\epsilon_{cs}$ per unit length as in Figure 6.6(a). If, however, tensile forces N_{cs} were then applied to each end, as in Figure 6.6(b) acting at the centroid of the slab and parallel to the axis of the beam so as to restore the slab to the length of the precast element and bring about an undeformed condition of the composite member, a uniform tensile stress $\Delta\epsilon_{cs}E_c$ would be induced in the *in situ* concrete, and the magnitude of the force required would be

$$N_{cs} = \Delta\epsilon_{cs}E_cA_i$$

where A_i = area of *in situ* concrete

The composite member must, however, be in a state of internal equilibrium with no external forces acting. The tensile forces N_{cs} must therefore be removed, which is equivalent to the application of equal and opposite compressive forces along the same line. The compressive forces will induce direct and bending stresses in the composite section [Figure 6.6(c)] which, in the *in situ* concrete, must be imposed on the existing tensile stress.

In the previously worked numerical example, for which the dimensions of the section are given in Figure 6.3, if the differential shrinkage is 100×10^{-6} and $E_c = 30 \ \mathrm{kN/mm^2}$

$$\Delta\epsilon_{cs}E_c = 100 \times 10^{-6} \times 30 \times 10^3 = 3{\cdot}00 \ \mathrm{N/mm^2}$$

$$N_{cs} = \Delta\epsilon_{cs}E_cA_i = 3{\cdot}00 \times 750 \times 120/10^3 = 270 \ \mathrm{kN}$$

Eccentricity of force N_{cs} from centroid of composite section,

$$y = 620 - 250 - 180 - 120/2 = 130 \ \mathrm{mm}$$

Moment $N_{cs}y = 270 \times 130/10^3 = 35{\cdot}1 \ \mathrm{kNm}.$

Differential shrinkage stress in beam (tensile stresses positive):

$$f_{inf} = -270 \times 10^3/155 \ 000 + 35{\cdot}1/13{\cdot}0 = +0{\cdot}96 \ \mathrm{N/mm^2}$$

$$f_{sup} = -270 \times 10^3/155 \ 000 \pm 35{\cdot}1/79{\cdot}8 = -2{\cdot}18 \ \mathrm{N/mm^2}$$

Differential stresses in slab:

$$f_{inf} = -2{\cdot}18 + 3{\cdot}00 = +0{\cdot}82 \ \mathrm{N/mm^2}$$

$$f_{sup} = -270 \times 10^3/155 \ 000 - 35{\cdot}1/29{\cdot}4 + 3{\cdot}00 = +0{\cdot}07 \ \mathrm{N/mm^2}$$

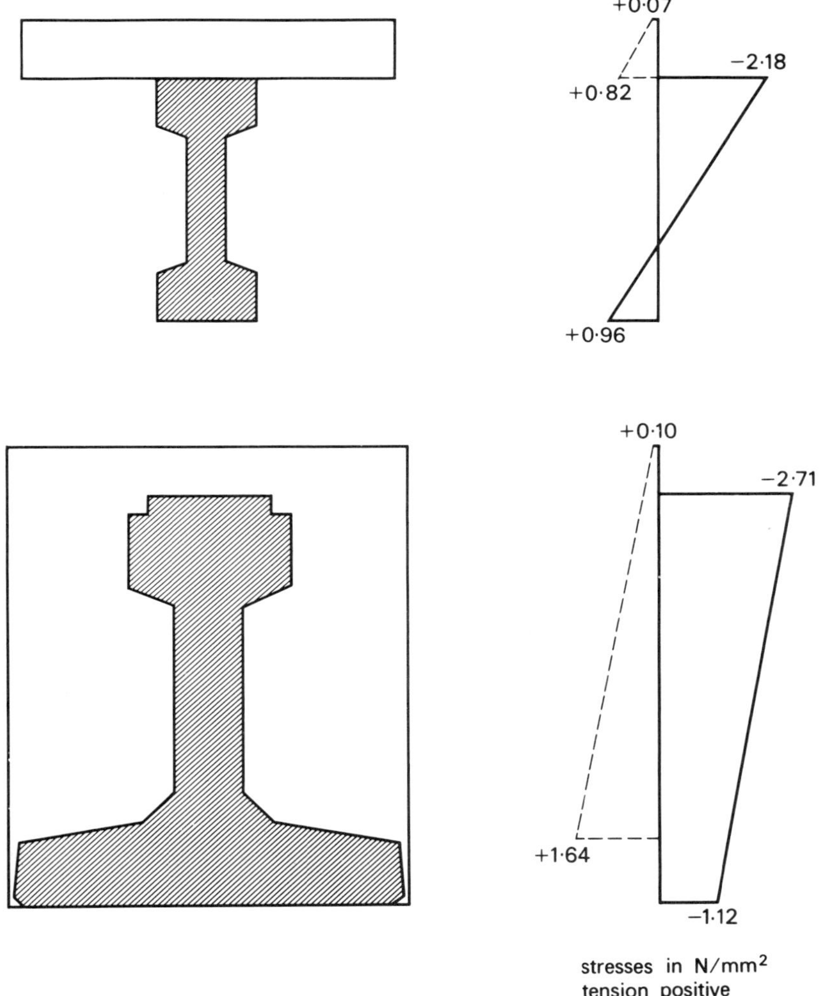

Figure 6.7 Examples of stresses induced by a differential shrinkage of 100×10^{-6}

The same method may be used to calculate the stresses due to differential shrinkage in a composite slab with inverted T beams. This has been done for the example used earlier in this chapter [Figure 6.4(a)] and the stress distribution diagrams for the two types of section are compared in Figure 6.7. In the beam and slab a tensile stress is induced at the soffit of the prestressed beams which would cause a slight reduction of the cracking load. On the other hand, the lower centroid of the *in situ* concrete in the composite slab results in compressive stress throughout the prestressed beam and tensile stress in the *in situ* concrete. It is unlikely that the tensile stress at the interface with the bottom flange of the beam would materially affect the behaviour having regard to the restraint of the uncracked flange, as discussed earlier. In most instances the effect of creep would be to relieve some of the stress calculated assuming a constant modulus of elasticity.

6.8 DIMENSIONING OF COMPOSITE SECTIONS

Formulae may be derived which relate the section moduli of the precast prestressed section and the composite section, given the loading, the allowable stresses in the concrete, and the loss ratio. Such formulae afford a guide for the dimensioning of the sections or for the choice of standard sections when these are to be used, and eliminate repeated analysis of the stresses to arrive at the correct size of section. The method of derivation is analogous to that used in Chapter 4.

The critical stress condition is normally at the soffit of the precast prestressed beam. Before placing the *in situ* concrete, when the minimum moment M_{min} (normally due to the self-weight of the beam) is acting:

$$f_{inf} - \frac{M_{min}}{Z_{1\,inf}} \leqslant f_{cp\,adm} \tag{6.1}$$

After the maximum loss of prestress, denoted by the ratio η and when the full service load $M_1 + M_2$ is applied:

$$\frac{M_1}{Z_{1\,inf}} + \frac{M_2}{Z_{2p\,inf}} - \eta f_{inf} \leqslant f_{cp\,adm} \tag{6.2}$$

In the above expressions the allowable stresses are as defined in Chapter 4. Eliminating the prestress f_{inf} from (6.1) and (6.2)

$$Z_{2p} \geqslant \frac{M_2 Z_{1\,inf}}{Z_{1\,inf}(\eta f_{cp\,adm} + f_{t\,adm}) - (M_1 - \eta M_{min})} \tag{6.3}$$

Formula (6.3) enables the required size of composite section to be determined, using a selected size of precast section. The method is illustrated in the following example.

Example

Calculate the required depth of a composite slab, incorporating the C and CA standard inverted T beam M_1, ref. (6.4), to support a characteristic imposed load of 15 kN/m² on a span of 15 m. Determine the minimum prestress necessary.

The beam will be designed with straight parallel pretensioned tendons assuming a loss ratio of 0·75, using Grade 40 concrete and specifying a strength of 35 N/mm² at transfer. Grade 30 concrete will be specified for the *in situ* part. With straight tendons $M_{min} = 0$ (at the support). The depth of the slab is estimated as 800 mm to obtain the load due to the weight of the beam and *in situ* concrete,

$$g_1 = 1 \cdot \dot{0} \times 0 \cdot 8 \times 2400 \times 9 \cdot 81/10^3 = 19 \text{ kN/m}$$

$$M_1 = 19 \times 15^2 \times 0 \cdot 125 = 534 \text{ kNm}$$

$$M_2 = 15 \times 15^2 \times 0 \cdot 125 = 422 \text{ kNm}$$

The permissible stresses are

$$f_{cp \text{ adm}} = 0 \cdot 5 \times 35 = 17 \cdot 5 \text{ N/mm}^2$$

$$f_{t \text{ adm}} \text{ (CP 110, Table 33)} = 2 \cdot 9 \text{ N/mm}^2$$

For beam M_1, $Z_{1 \text{ inf}} = 47 \cdot 17 \text{ N/mm}^2$
From formula (6.3)

$$Z_{2p \text{ inf}} \geqslant \frac{422 \times 47 \cdot 17 \times 10^6}{47 \cdot 17(0 \cdot 75 \times 17 \cdot 5 + 2 \cdot 9) - (534 - 0)}$$

$$= 89 \cdot 71 \times 10^6 \text{ mm}^3$$

If h_2 = total depth of slab,

$$1000 h_2{}^2/6 = 89 \cdot 71 \times 10^6$$

$$h_2 = \left(\frac{6 \times 89 \cdot 71 \times 10^6}{1000}\right)^{\frac{1}{2}} = 734 \text{ mm}$$

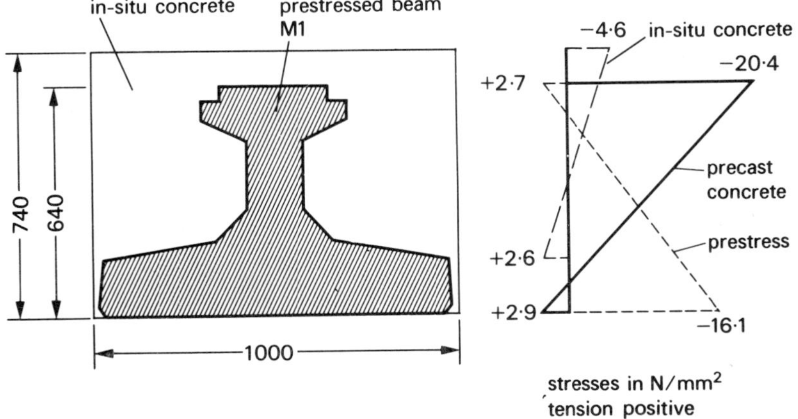

Figure 6.8 Design of composite slab for a standard prestressed beam

A depth of 740 mm will be adopted (Figure 6.8); the corrected value of M_1 is then 490 kNm, and $Z_{2p\ inf} = 1000 \times 740^2/6 = 91\cdot27 \times 10^6$ mm³.

The minimum prestress, f_{inf}, required at the soffit of the precast beam is calculated from a rearrangement of formula 6.1:

$$f_{inf} \geqslant \frac{1}{\eta}\left(\frac{M_1}{Z_{1\ inf}} + \frac{M_2}{Z_{2p\ inf}} - f_{t\ adm}\right)$$

$$= \frac{1}{0\cdot75}\left(\frac{490}{47\cdot17} + \frac{422}{91\cdot27} - 2\cdot9\right) = 16\cdot1\ \text{N/mm}^2$$

The maximum tensile prestress, f_{sup}, required at the top of the precast beam is obtained from formula 4.2 as explained in Chapter 4 (p. 52–53).

$$f_{sup} \geqslant \frac{M_{min}}{Z_{1\ sup}} + f_{tp\ adm}$$

$$= 0 + 2\cdot7 = 2\cdot7\ \text{N/mm}^2\ \text{(tensile)}$$

A suitable arrangement of tendons may now be designed by the method described in Chapter 4.

The stresses at the various points in the composite section are checked, and the cross-section and stress distribution diagram are shown in Figure 6.8 It will be noted that the tensile stress in the *in situ* concrete at the interface with the bottom flange of the prestressed beam is less than the allowable value of 3·6 N/mm² for Grade 30 concrete (CP 110, Table 44) and that the compressive stress at the top of the prestressed beam is about equal to the allowable value of $0\cdot33 \times 40 \times 1\cdot5 = 20$ N/mm².

Formula (6.3) may be arranged in an alternative form to indicate the required section modulus of the precast prestressed section in a composite slab of given depth.

$$Z_{1\ inf} \geqslant \frac{Z_{2p\ inf}(M_1 - \eta M_{min})}{Z_{2p}(\eta f_{cp\ adm} + f_{t\ adm}) - M_2} \tag{6.4}$$

Example
Calculate the size of a precast beam for a composite slab of total depth 1050 mm *for an imposed service load of* 15 kN/m² *on a span of* 20 m *and the minimum prestress required.*

The same grades of concrete will be specified as for the previous example. The longer span beams, however, will be designed to have deflected tendons; the minimum moment will therefore be taken as the moment at mid-span due to the self-weight of the beam.

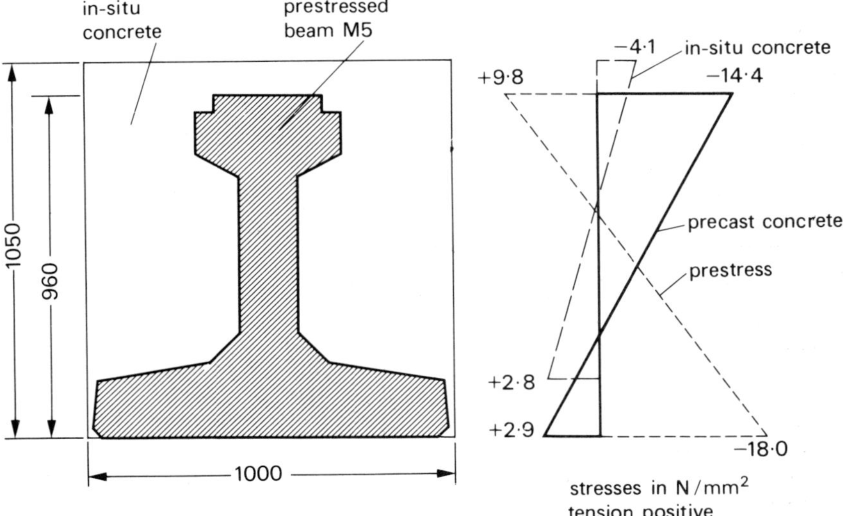

Figure 6.9 Selection of standard prestressed beam for a composite slab of given depth

Assuming an M5 beam for calculation of the self-weight

$$g_{\min} = 8\cdot 37 \text{ kN/m}$$

Load due to weight of beam and *in situ* concrete,

$$g_1 = 1\cdot 0 \times 1\cdot 05 \times 2400 \times 9\cdot 81/10^3 = 24\cdot 72 \text{ kN/m}$$

$$M_{\min} = 8\cdot 37 \times 20^2 \times 0\cdot 125 = 418 \text{ kNm}$$

$$M_1 = 24\cdot 72 \times 20^2 \times 0\cdot 125 = 1236 \text{ kNm}$$

$$M_2 = 15 \times 20^2 \times 0\cdot 125 = 750 \text{ kNm}$$

$$Z_{2p\,\inf} = 1000 \times 1050^2/6 = 183\cdot 75 \times 10^6 \text{ mm}^3$$

From formula (6.4)

$$Z_{1\,\inf} \geqslant \frac{183\cdot 75(1236 - 0\cdot 75 \times 418) \times 10^6}{183\cdot 75(0\cdot 75 \times 17\cdot 5 + 2\cdot 9) - 1236} = 99\cdot 21 \times 10^6 \text{ mm}^3$$

A beam M_1 which has the section moduli $Z_{1\,\inf} = 100\cdot 33 \times 10^6$ mm³, $Z_{1\,\sup} = 59\cdot 39 \times 10^6$ mm³, is therefore adequate.

The prestress is calculated and the stresses checked as in the previous example. The cross section and stress distribution is shown in Figure 6.9.

The detailing of C and CA standard inverted T prestressed beams and composite slabs has been considered in depth in a recent publication ref. (6.8).

E

References

6.1 MORICE, P. B. and LITTLE, G., 'The analysis of right bridge decks subjected to abnormal loading,' Publication Db 11, Cement and Concrete Association, London, 1956.

6.2 ROWE, R. E., *Concrete Bridge Design*, CR Books, London, 1962, p. 336.

6.3 SOMERVILLE, G. and TILLER, R. M., 'Standard bridge beams for spans from 7 m to 36 m', Publication 32.005 Cement and Concrete Association, London, 1970.

6.4 MANTON, B. H. and WILSON, C. B., 'MoT/C&CA standard bridge beams', Publication 32.012 Cement and Concrete Association, London, 1970, p. 36.

6.5 WEST, R., 'Recommendations on the use of grillage analysis for slab and pseudo-slab bridge decks', Departmental Note ITN 7003, Cement and Concrete Association, London, 1970.

6.6 MINISTRY OF TRANSPORT AND CONSTRUCTION INDUSTRY RESEARCH AND INFORMATION ASSOCIATION, 'Finite element package for analysis of reinforced concrete slab bridge decks', Program BECP/1, London, 1969.

6.7 CHAPLIN, E. C. *et al.*, 'The development of a design for a precast concrete bridge beam of U section', *Structural Engineer*, **51**, 10, October, 1973.

6.8 GREEN, J. K., 'Detailing for standard prestressed concrete bridge beams', Publication 46.018, Cement and Concrete Association, London, 1973.

7

Statically Indeterminate Structures

7.1 REDUNDANT REACTIONS AND SECONDARY MOMENTS

Statically indeterminate prestressed concrete structures have certain advantages, but also introduce a number of problems. Continuity of the members in a framed structure leads to increased stability, and the connection of abutting members to form an indeterminate structure by prestressing the contact surfaces offers one possible solution to the notorious problem of the instability of structures assembled from precast concrete elements. The distribution of moments in a statically indeterminate structure usually permits a small reduction in the sizes of the members, and a further economic advantage of a continuous prestressed beam over a series of simply supported beams lies in the fact that one pair of post-tensioning anchorages and a single stressing operation can serve several members.

One of the problems associated with indeterminate prestressed structures is the greater amount of curvature of the tendons which is often necessary in order to locate the prestressing force correctly; this leads to a large loss of prestress due to friction, as discussed in Chapter 2. The major difficulty, however, is the greater complexity of the design calculations on account of the secondary effects caused by the prestressing of an indeterminate structure. Most of this chapter is devoted to the theory of secondary moments.

Secondary moments are generated in a statically indeterminate structure by the additional redundant reactions induced by the deformation of the structure under the prestress. A statically determinate structure is free to deform when prestressed, but in an indeterminate structure the restraint of the redundancies, whether these be supports or members, causes reactions. The principle is illustrated by the simple example of a continuous beam resting

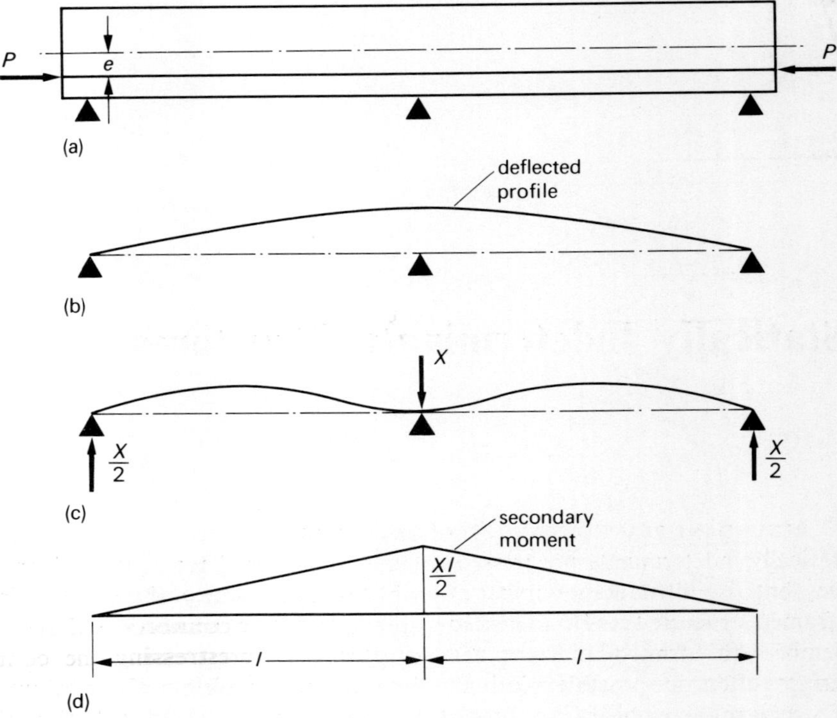

Figure 7.1 Redundant reactions and secondary moment in a continuous beam

on three supports as in Figure 7.1(a). If the beam is prestressed by a straight
tendon parallel to the axis it will, unless restrained at the central support,
assume an upward camber under the bending action of the prestressing force
[Figure 7.1(b)]. In order to maintain the centre of the beam at the level of the
central support a reaction, downward in this instance, must be exerted at the
support [Figure 7.1(c)]. The action of prestressing will therefore induce a
secondary moment [Figure 7.1(d)] in addition to the primary moment (i.e.
the prestressing force multiplied by its eccentricity).

The total effect of the prestressing force, including the secondary effect,
may be conveniently represented by the resultant line of thrust through the
beam. The diagram of the resultant bending moment $M - Pe$ of the prestressing moment and secondary moment in the above example is shown in
Figure 7.2. In order to produce this moment the prestressing force P would
have to act with an eccentricity equal to $M/P - e$. The resultant line of thrust
is thus the line of the prestressing force (i.e. the tendon profile) vertically
displaced a distance M/P, where M is the secondary moment.

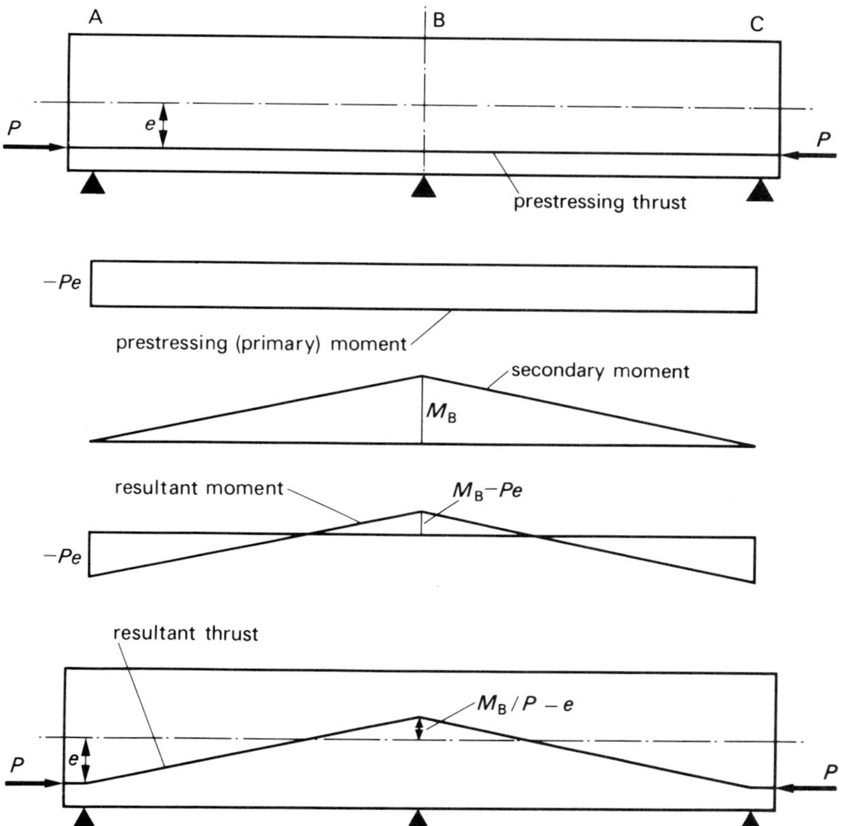

Figure 7.2 Resultant thrust in a continuous beam

7.2 ANALYSIS OF SECONDARY MOMENTS BY MOMENT DISTRIBUTION

In principle, any method of linear analysis may be used to calculate the secondary moments in a statically indeterminate prestressed structure. Whatever method is applied it will be based on the fact that the free bending moment at each point will be equal to the Pe prestressing force multiplied by its eccentricity at that point. The free bending moment diagram for each member will therefore be represented by the tendon profile, with the longitudinal axis of the member as the horizontal axis and one vertical unit of distance being equal to P units of moment. Sagging moments (or more generally moments inducing concave curvature with reference to the centre of a frame) will occur when the tendon is above the axis (or away from the centre of the frame) and will be taken as positive.

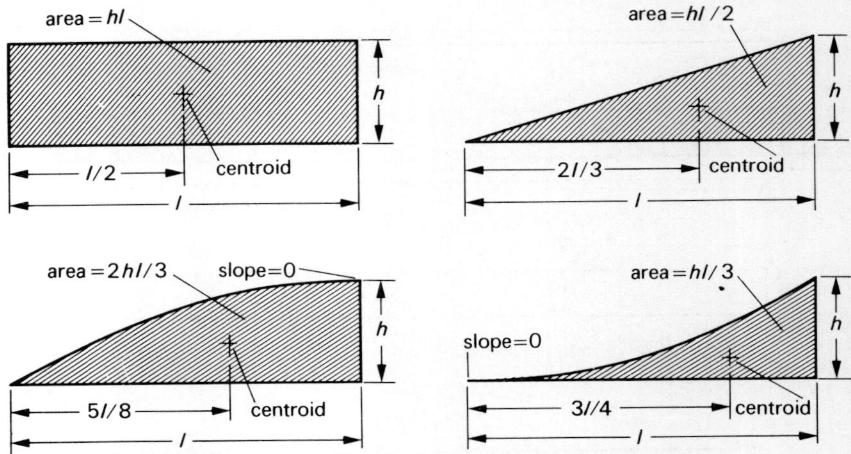

Figure 7.3 Basic shapes for area-moment calculations

Curved lengths of cable may be considered to be parabolic, and the area and area-moment for a complex tendon profile may be treated as the sum of a number of positive and negative areas of the four basic shapes shown in Figure 7.3. The technique will be clear from the following examples, in which the moment distribution method has been used as it is probably the method most widely adopted for the analysis of small structures using a slide rule or desk calculating machine.

Example

Calculate the secondary moment in the two-span continuous prestressed concrete beam ABA' illustrated in Figure 7.4 and draw the resultant line of thrust. The inflected tendon profile in each span consists of two equal lengths of equal and opposite curvature.

Figure 7.4 shows how the area enclosed between the tendon profile and the axis is analysed into areas of simple shape. The dimensions and distance of the centroid from the left-hand end of the member are shown for each of the component areas, and the vertical dimension will determine whether the area and its moment are positive or negative.

The fixed-end moments due to prestressing may now be calculated. Referring to Figure 7.5, these are given by the expressions:

$$M_{\mathrm{FA}} = -\frac{6A_{\mathrm{M}}}{l^2}\left(\frac{2l}{3} - \bar{x}\right)$$

$$M_{\mathrm{FB}} = +\frac{6A_{\mathrm{M}}}{l^2}\left(\bar{x} - \frac{l}{3}\right)$$

where A_{M} = area of the prestressing moment

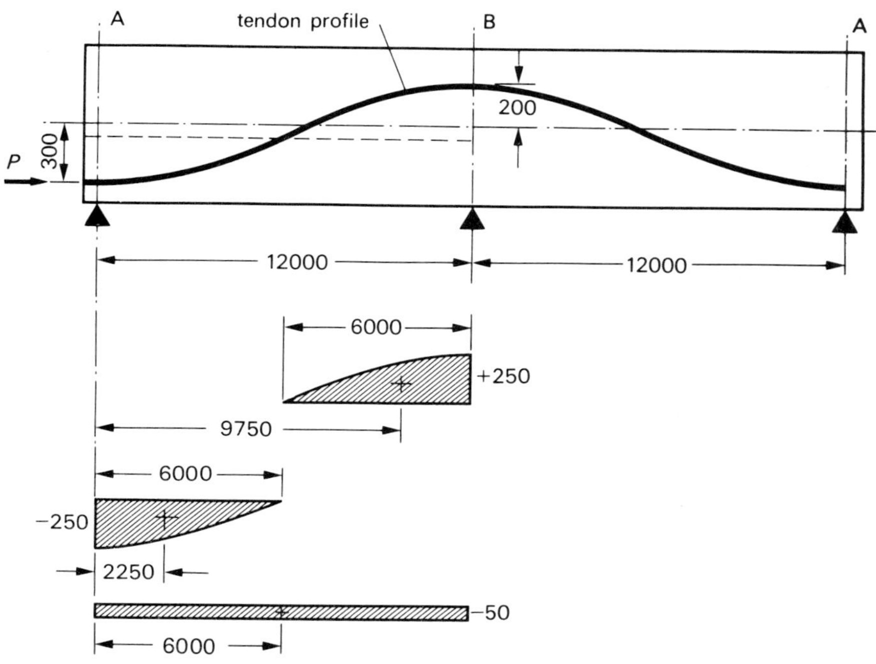

Figure 7.4 Analysis of area under tendon profile of two-span continuous beam

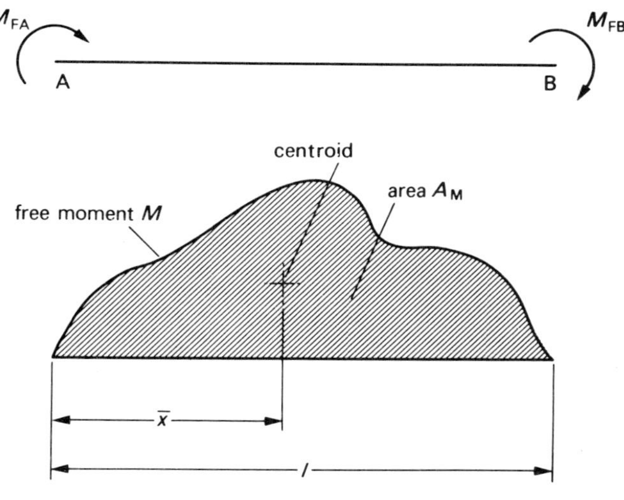

Figure 7.5 Fixed end moments

$$(= P \times \text{ area under tendon profile})$$

$$\bar{x} = \text{distance of centroid of area } A_M \text{ from left-hand end of member}$$

$$l = \text{length of member}$$

It must be remembered that a different sign convention is necessary for the moment distribution method, whereby moments are defined as positive when acting in a clockwise direction on the end of the member.

In the present example, the fixed-end moments are calculated as follows, putting horizontal lengths in metres and vertical lengths in millimetres, so that the computed moments are in millimetres × units of force:

$$M_{FA} = -\frac{6P}{12^2}[(2/3) \times 250 \times 6(8 \cdot 00 - 9 \cdot 75) - (2/3) \times 250$$

$$\times 6(8 \cdot 00 \times 2 \cdot 25) - 50 \times 12(8 \cdot 00 - 6 \cdot 00)]$$

$$= +362P \text{ (mm} \times \text{units of force } P)$$

$$M_{FB} = +\frac{6P}{12^2}[(2/3) \times 250 \times 6(9 \cdot 75 - 4 \cdot 00) - (2/3) \times 250$$

$$\times 6(2 \cdot 25 - 4 \cdot 00) - 50 \times 12(6 \cdot 00 - 4 \cdot 00)]$$

$$= +262P \text{ (mm} \times \text{units of force } P)$$

In this very simple symmetrical indeterminate structure, with only one degree of indeterminacy, all that is required is a single carry-over operation, which is identical in the two spans as shown in Figure 7.6. The secondary moment at the central support B is found to be $+81P$ mm × units of force.

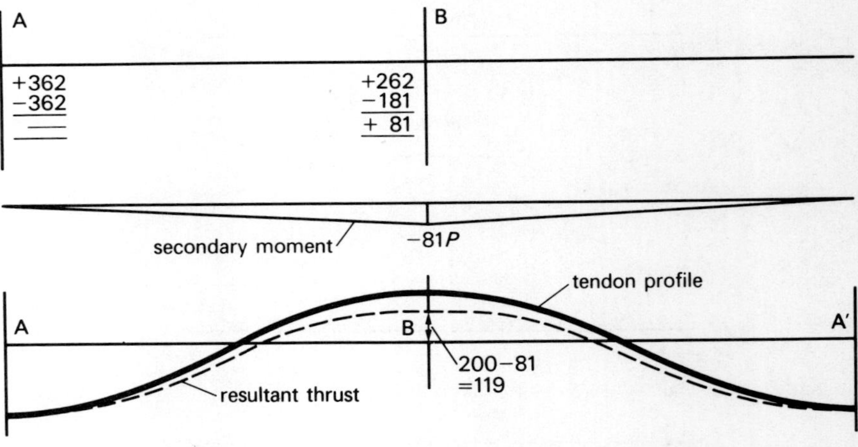

Figure 7.6 Calculation of secondary moment and resultant thrust by moment distribution

According to the moment distribution sign convention this is a moment acting in the clockwise sense on the end of the member AB and is therefore a hogging moment which is negative according to the sign convention adopted for the free prestressing moment and line of thrust. The secondary moment therefore varies linearly from zero at the outer supports to a value of $-81P$ at the central support at which point the effect is to displace the line of thrust downwards a distance of 81 mm. The diagrams of the secondary moment, the tendon profile and the resultant line of thrust are given in Figure 7.6 from which it may be seen that the intrinsic shape of the resultant line of thrust is the same as that of the tendon profile, but that the secondary moment has brought about a translation and rotation.

Example

A further example of the calculation of secondary moments by the method of moment distribution is provided by the symmetrical fixed portal in Figure 7.7, in which the transom is prestressed by a parabolic tendon and each leg by a straight tendon.

The breakdown of the area between the tendon profile and the axis of each member is shown in Figure 7.7. The fixed-end moments are computed as in the previous example.

$$M_{\text{FAB}} = -\frac{6A_{\text{M}}}{l^2}\left(\frac{2l}{3} - \bar{x}\right)$$

$$= -\frac{6P_{\text{AB}}}{5^2}\left[(1/2) \times 100 \times 5\cdot000(3\cdot333 - 3\cdot333) - 50\right.$$

$$\left. \times 5\cdot000(3\cdot333 - 2\cdot500)\right]$$

$$= +50P_{\text{AB}} = 8\cdot000 \text{ kNm}$$

$$M_{\text{FBA}} = +\frac{6A_{\text{M}}}{l^2}\left(\bar{x} - \frac{l}{3}\right)$$

$$= +\frac{6P_{\text{AB}}}{5^2}\left[(1/2) \times 100 \times 5\cdot000(3\cdot333 - 1\cdot667) - 50\right.$$

$$\left. \times 5\cdot000(2\cdot5000 - 1\cdot667)\right]$$

$$= +50P_{\text{AB}} = +8\cdot000 \text{ kNm}$$

$$M_{\text{FBB}'} = -\frac{6P_{\text{BB}}}{6^2}\left[50 \times 6\cdot000(4\cdot000 - 3\cdot000) - (2/3) \times 150\right.$$

$$\left. \times 6\cdot000(4\cdot000 - 3\cdot000)\right]$$

$$= +50P_{\text{BB}'} = +7\cdot000 \text{ kNm}$$

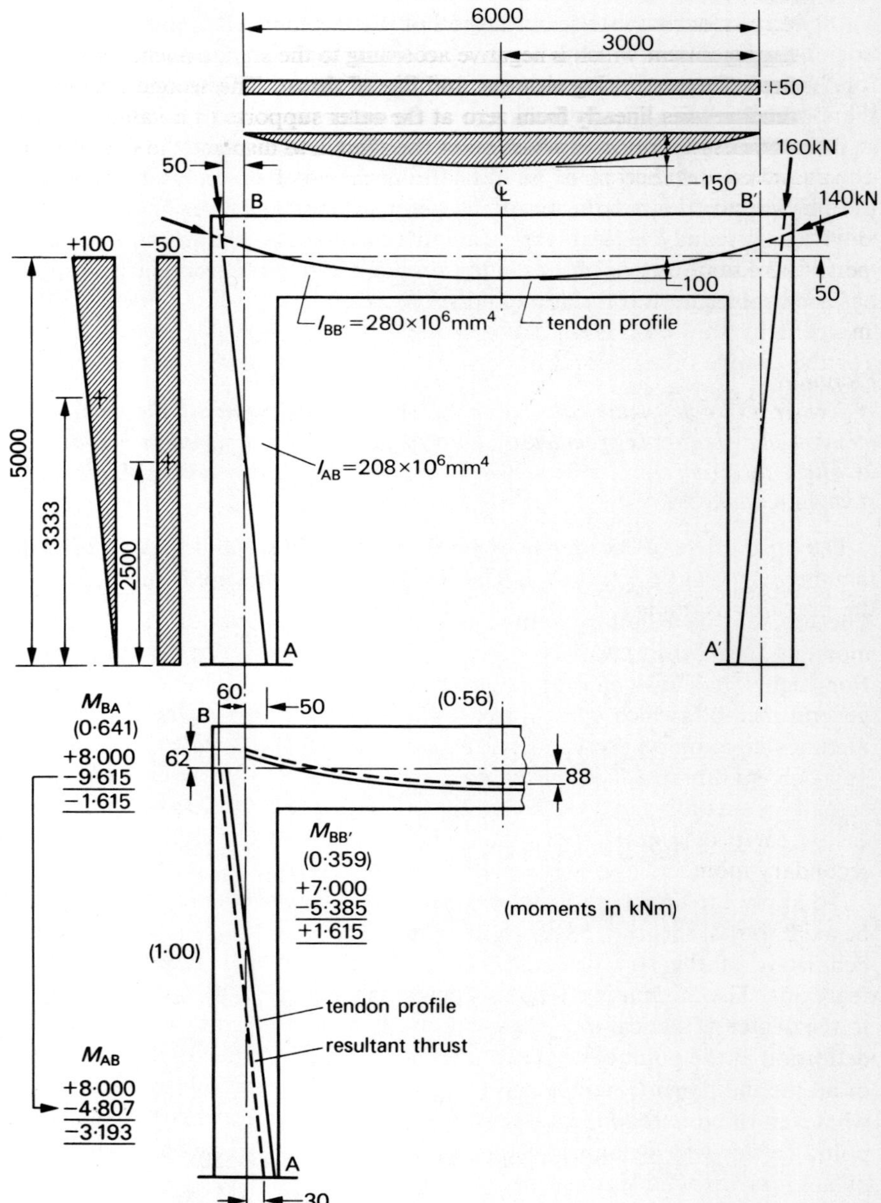

Figure 7.7 Secondary moments and resultant thrust in a fixed portal

$$\text{Stiffness ratio of BB' to AB} = \frac{I_{BB'}}{l_{BB'}} \times \frac{l_{AB}}{I_{AB}}$$

$$= \frac{280 \times 10^6}{6000} \times \frac{5000}{208 \times 10^6} = 1\cdot12 : 1\cdot00$$

The stiffness of BB' is halved, reducing the ratio to $0\cdot56 : 1\cdot00$ to enable the moment distribution to be carried out in one-half of the symmetrical frame. The distribution factors at the joint B are then $1\cdot00/(1\cdot00+0\cdot56) = 0\cdot641$ in BA and $0\cdot359$ in BB', and the distribution operations are easily performed as in Figure 7.7.

The displacement of the line of thrust is obtained by dividing the secondary moment by the prestressing force in each member, making due allowance for the change of sign convention, and the eccentricity of the resultant line of thrust is as follows:

$$e_{AB} = -50 + 3193/160 = -30 \text{ mm}$$

$$e_{BA} = +50 + 1615/160 = +60 \text{ mm}$$

$$e_{BB'} = +50 + 1615/140 = +62 \text{ mm}$$

7.3 ANALYSIS OF SECONDARY MOMENTS BY FLEXIBILITY INFLUENCE COEFFICIENTS

The moment distribution method is suitable for the analysis of secondary moments in structures with a moderately low degree of statical indeterminacy. For highly indeterminate structures, however, it is desirable to adopt a general method, which can be used on a computer. One of the two general methods in common use is the stiffness method, also known as the displacement or equilibrium method, of which the moment distribution method is actually a special variation. The other is the flexibility method, also known as the force or compatability method, the use of which to calculate the secondary moments in a prestressed structure will now be considered.

To apply the flexibility method internal or external reactions, which may be axial forces, shearing forces or moments, must first be inserted so that the behaviour of the structure can be completely defined in terms of these reactions. The minimum number of reactions required for this will be equal to the degree of statical indeterminacy of the structure. An expression for the deflection at the point of application of each reaction is then written in terms of all the unknown reactions and the external load and is equated to zero or whatever value is required to satisfy the condition of compatability at this point. In this way a number of simultaneous equations equal to the number of unknown reactions is set up, enabling the reactions to be solved.

The equations involve flexibility influence coefficients, defined as follows:

$$f_{ij} = \text{deflection at the point } i \text{ caused by a unit reaction at the point } j$$

u_i = deflection at the point i due to the external load on the structure

If R_1, R_2, ... R_n, are the unknown reactions at the points 1, 2, ... n, the compatability equations will be:

$$f_{11}R_1 + f_{12}R_2 + \ldots f_{1n}R_n + u_1 = 0 \qquad \ldots 1$$

$$f_{21}R_1 + f_{22}R_2 + \ldots f_{2n}R_n + u_2 = 0 \qquad \ldots 2$$

$$f_{n1}R_1 + f_{n2}R_2 + \ldots f_{nn}R_n + u_n = 0 \qquad \ldots n$$

The n equations are in a suitable form for solution by computer, and a matrix formulation may be used if desired.

The flexibility influence coefficients required for the above equations are given by the expressions

$$f_{ij} = \int \frac{m_i m_j}{EI} \mathrm{d}s$$

$$u_i = \int \frac{m_0 m_i}{EI} \mathrm{d}s$$

where m_i = moment induced by a unit reaction at the point i
        m_j = moment induced by a unit reaction at the point j
        m_0 = moment due to external load on the structure
        $\mathrm{d}s$ = element of length of a member; the integration is performed for
              all the members in the structure

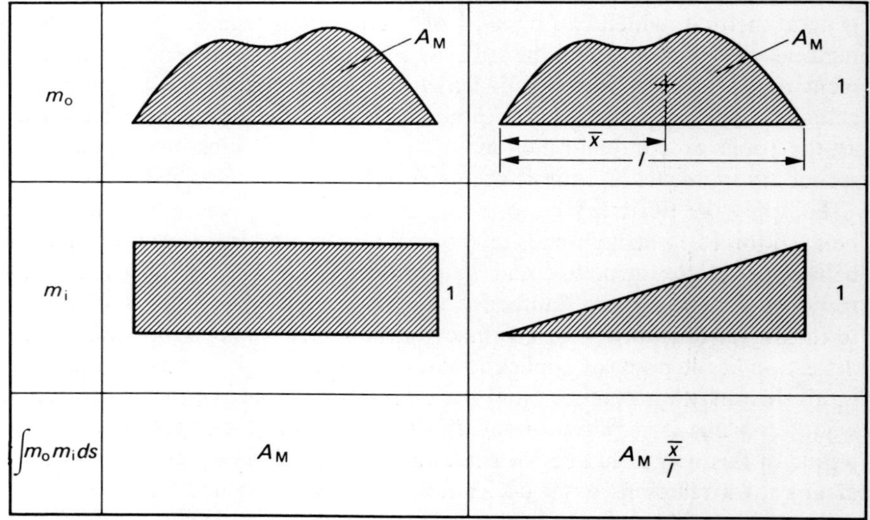

Figure 7.8 Product integrals for flexibility influence coefficients

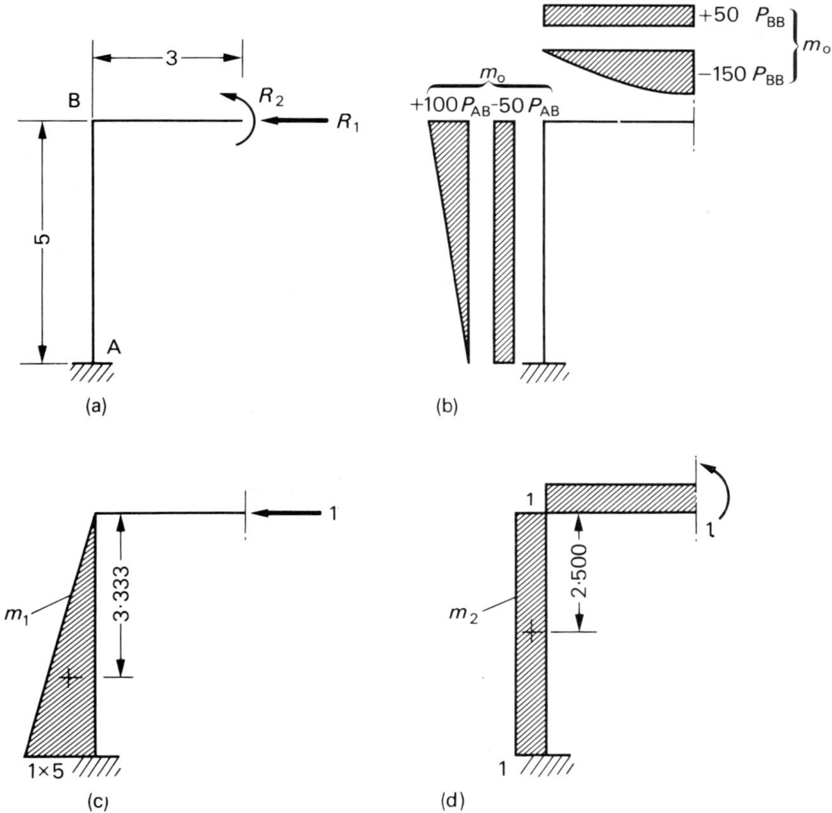

Figure 7.9 Calculation of secondary moment in a fixed portal by flexibility influence coefficients

Tables are provided of the product integrals $m_i m_j$ and $m_0 m_i$ for the common shapes of bending moment diagram in most textbooks on structural analysis. In the calculation of secondary moments due to prestressing, m_0 is the bending moment applied by the eccentric prestressing force, and the integral of this moment is equal to the product of the prestressing force and the area intercepted between the tendon profile and the axis. The moment m_i induced by a unit reaction is usually linear, and the product integral $m_0 m_i$ for each span can easily be evaluated as shown in Figure 7.8, dividing the area between the tendon profile and the axis into simple components in the same way as when calculating the fixed-end moment for the moment distribution method. An example will make the method clear.

Example

Calculate the secondary moments in the fixed portal frame in Figure 7.7 by the flexibility method.

The loading due to the prestress is symmetrical and the structure therefore statically indeterminate to the second degree. Half of the frame may be analysed by inserting as the two redundants the thrust R_1 and the moment R_2 at the mid-span point of the transom, as in Figure 7.9(a). Figure 7.9(b) shows the moment applied by prestressing m_0 divided into basic components, and the moments m_1 and m_2 induced by unit reactions in the directions of R_1 and R_2 are shown in (c) and (d) respectively.

The flexibility coefficients are evaluated, using the metre as the unit of length, as follows:

$$f_{11} = \frac{1}{EI_{AB}} [(1/2) \times 5 \times 5 \times 3{\cdot}333] = 41{\cdot}667/EI_{AB}$$

$$f_{12} = f_{21} = \frac{1}{EI_{AB}} [(1/2) \times 5 \times 5] = 12{\cdot}500/EI_{AB}$$

$$f_{22} = \frac{1}{EI_{AB}} [1 \times 5] + \frac{1}{EI_{BB'}} [1 \times 3]$$

$$= \frac{1}{EI_{AB}} [5 + (208/280) \times 3] = 7{\cdot}229/EI_{AB}$$

$$u_1 = \frac{160}{EI_{AB}} [(1/2) \times 0{\cdot}10 \times 5 \times 5 \times 1{\cdot}667 - 0{\cdot}05 \times 5 \times 2{\cdot}500]$$

$$= -33{\cdot}333/EI_{AB}$$

$$u_2 = \frac{160}{EI_{AB}} [(1/2) \times 0{\cdot}10 \times 5 - 0{\cdot}05 \times 5]$$

$$+ \frac{140}{EI_{BB'}} [0{\cdot}05 \times 3 - (2/3) \times 0{\cdot}15 \times 3]$$

$$= -15{\cdot}600/EI_{AB}$$

The compatability equations may now be written, cancelling the factor $1/EI_{AB}$ throughout.

$$41{\cdot}667R_1 + 12{\cdot}500R_2 - 33{\cdot}333 = 0$$

$$12{\cdot}500R_1 + 7{\cdot}229R_2 - 15{\cdot}600 = 0$$

Hence,

$$R_1 = 0{\cdot}317 \text{ kN}$$

$$R_2 = 1{\cdot}610 \text{ kNm}$$

The secondary moments are therefore:

$$M_\mathrm{A} = 1 \cdot 610 + 0 \cdot 317 \times 5 \cdot 000$$

$$= 3 \cdot 195 \text{ kNm}$$

$$M_\mathrm{B} = 1 \cdot 610 \text{ kNm}$$

7.4 THE TENDON REACTION METHOD OF ANALYSIS

An alternative approach to the analysis of a statically indeterminate pre-stressed structure is to consider the various reactions exerted on the concrete by the system of tendons. At each end anchorage of a post-tensioned tendon there will be a concentrated force which may act eccentrically and at an angle to the axis of the member [Figure 7.10(a)]. There will also be transverse reactions at intermediate points on the tendon resulting from its bearing on the wall of the duct where it changes direction. Where the tendon is curved the reaction will be a distributed load [Figure 7.10(b)] and at a point where the direction of a tendon is sharply deflected there will be a concentrated load [Figure 7.10(c)].

Exact and approximate expressions for the above reactions are given in Figure 7.10. The approximate expressions, which in certain instances neglect the effect of the inclination of the tendon to the axis are sufficiently accurate except where the tendon is steeply inclined. Greater accuracy may be obtained, if required, by correcting the local value of P by the loss due to friction.

Before applying the method of tendon reactions to the analysis of a statically indeterminate structure, the concept is illustrated in an example of a simply supported beam in Figure 7.11. The beam, being statically determinate, is free to deform under the action of the prestress, and the system of tendon reactions must therefore be in equilibrium. The resultant of the tendon reactions, for example on one-half of the beam can be shown to be equal to a force P parallel to and $0 \cdot 2$ below the axis, as would be anticipated from the method of sections which is normally employed.

The exactly calculated reactions are given in Figure 7.11(a). Resolving horizontally, the components of the resultant at the mid-span point are:

$$\text{normal force} = 0 \cdot 9950P + 0 \cdot 00083P \times 6 = 1 \cdot 00000P$$

$$\begin{aligned}
\text{shear force} &= -0 \cdot 0995P + 0 \cdot 01642P \times 6 + 0 \cdot 00019P \times 3 \\
&\quad + (1/2) \times 0 \cdot 00019P \times 3 + (1/2) \times 0 \cdot 00006 \times 3 = 0 \cdot 0000
\end{aligned}$$

$$\begin{aligned}
\text{moment} &= 0 \cdot 0995P - 0 \cdot 0995P \times 6 + 0 \cdot 01642P \times 6 \times 3 \\
&\quad + 0 \cdot 00019P \times 3 \times 1 \cdot 5 + (1/2) \times 0 \cdot 00019P \times 3 \times 4 \\
&\quad + (1/2) \times 0 \cdot 00006P \times 3 \times 1 = 0 \cdot 1999P \simeq 0 \cdot 2P
\end{aligned}$$

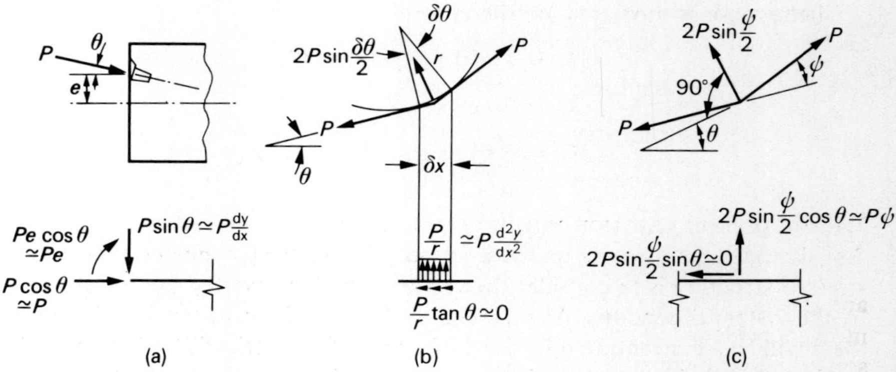

Figure 7.10 Reactions between tendon and concrete member: (a) end anchorage;
(b) curved tendon; (c) deflected tendon

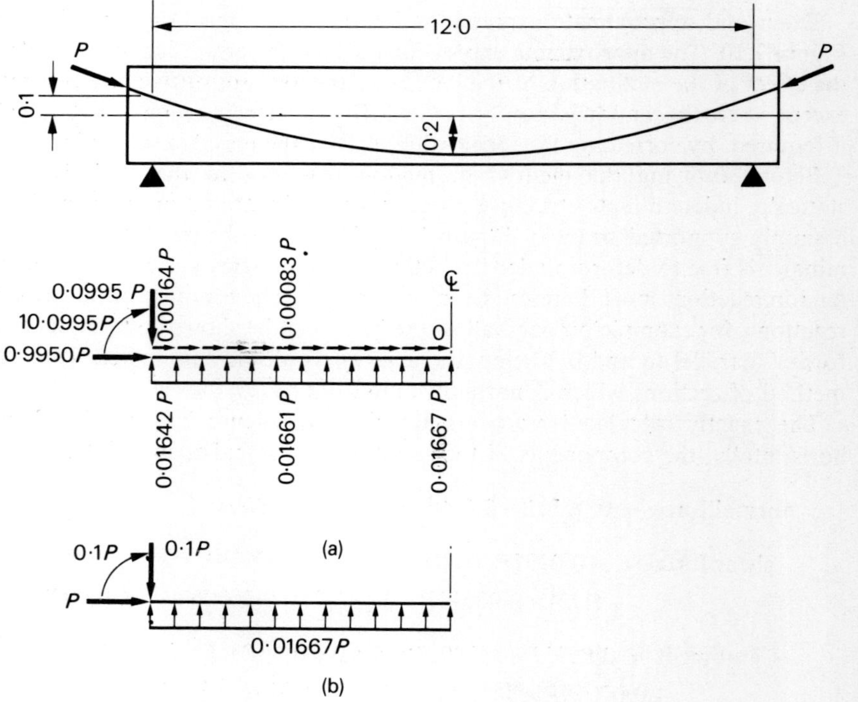

Figure 7.11 Reactions of post-tensioned tendon in a simply supported beam

However, a sufficiently accurate result may usually be obtained more rapidly from the approximate reactions, as given in Figure 7.11(b).

$$\text{normal force} = P$$

$$\text{shear force} = -0.1P - 0.01667P \times 6 = 0$$

$$\text{moment} = 0.1P - 0.1P \times 6 + 0.01667P \times 6 \times 3$$

$$= 0.2P$$

In the analysis of a statically indeterminate structure the tendon reactions are treated as externally applied loads and the terminal moment and span moments calculated by any method desired. The advantage of this approach sometimes known as the equivalent load method is that tables and formulae are available for the common systems of loading on a number of statically indeterminate structures.

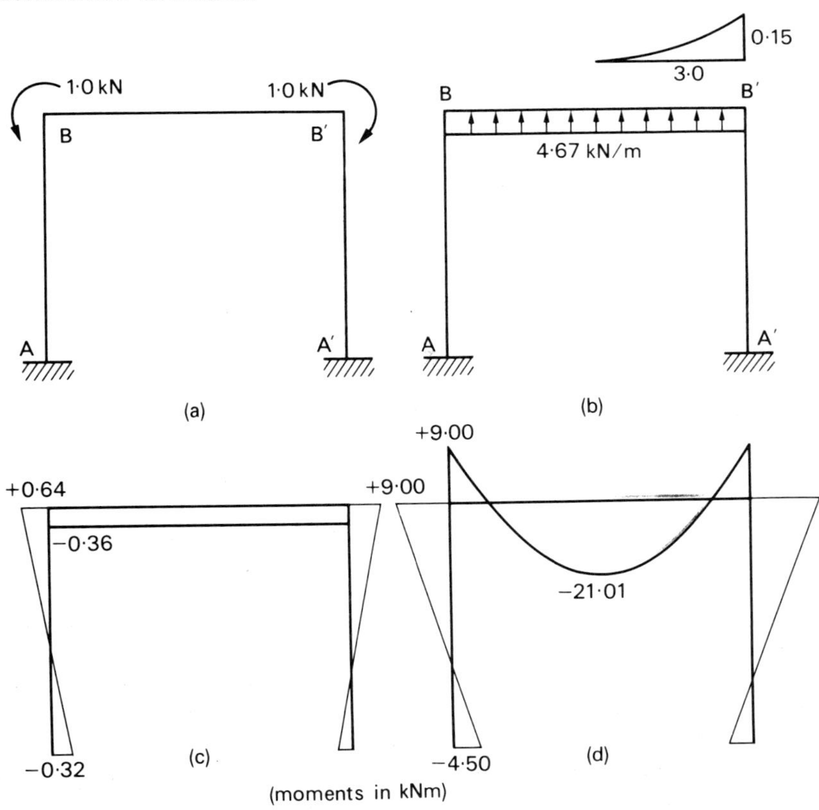

Figure 7.12 Analysis of fixed portal by tendon reaction method

It is important to note that the tendon reaction method yields the resultant effect of prestressing (i.e. prestressing moment + secondary moment).

Example

Use the tendon reaction method to analyse the prestressed fixed portal in the previous examples (Figure 7.7).

Using the approximate reactions (Figure 7.10), the combined effect of the two anchorages at B is an anticlockwise moment,

$$160 \times 0.05 - 140 \times 0.05 = 1.0 \text{ kNm} \quad [\text{Figure 7.12(a)}]$$

The moments induced by this and the symmetrically opposite couple at B′ are shown in Figure 7.12(c).

Since the tendons in AB and A′B′ are anchored in the fixed bases at A and A′, the only other significant tendon reaction is the upward distributed load from the curved tendon in BB′.

From the geometry of the parabola.

$$\frac{1}{r} \simeq \frac{\mathrm{d}^2 y}{\mathrm{d}x^2} = \frac{2 \times 0.15}{3.0^2} = 1/30 \text{ m} \quad [\text{Figure 7.12(b)}]$$

Distributed load $= P/r = 140/30 = 4.67$ kN/m.

The moment diagram resulting from this load is shown in Figure 7.12(d). The resultant moments are as follows:

$$M_{AB} = -0.32 - 4.50 = -4.82 \text{ kNm}$$

$$M_{BA} = +0.64 + 9.00 = +9.64 \text{ kNm}$$

$$M_{BB'} = -0.36 + 9.00 = +8.64 \text{ kNm}$$

The eccentricity of the resultant line of thrust at the ends of the members is therefore:

$$e_{AB} = -4.82 \times 1000/160 = +30 \text{ mm}$$

$$e_{BA} = +9.64 \times 1000/160 = +60 \text{ mm}$$

$$e_{BB'} = +8.64 \times 1000/140 = +62 \text{ mm}$$

A comparison will reveal that these are identical to the values obtained earlier by calculation of the secondary moments.

A special application of the tendon reaction method is in the design of statically indeterminate structures by 'load balancing', whereby the tendons are arranged so that the reactions exactly oppose the dead load and part of the imposed load. At this stage, therefore, the members of the structure will be in axial compression and the remainder of the imposed load must not exceed the flexural capacity of the structure.

7.5 THE CONCORDANT TENDON

It has been shown how, in general, the prestressing of a statically indeterminate structure gives rise to a sysem of secondary reactions and moments which result in a modification of the line of action of the prestressing force. It is possible, however, to arrange the tendon profile in such a way that the structure has no tendency to deform at the supports or other points of restraint, so that no redundant reactions or secondary moments are induced by prestressing. In this special case the tendon is said to be concordant.

The profile of a concordant tendon (or set of tendons) must fulfil a group of geometrical conditions, equal in number to the degree of indeterminacy of the structure; these conditions can usually be established from Mohr's two moment-area theorems. In the general case of a continuous beam with no settlement of the supports, the requirement of continuity of direction of the beam in the span n and the span $n+1$ (adjacent on the right) at the intervening support, yields the condition

$$\frac{A_{Mn}\bar{x}_n}{(EI)_n l_n} + \frac{A_{M\,n+1}(l_{n+1}-\bar{x}_{n+1})}{(EI)_{n+1} l_{n+1}} = 0$$

where A_{Mn} = area of free bending moment diagram for span n = prestressing force multiplied by area between tendon profile and axis of beam

     l_n = length of span n

     $\bar{x}_n$ = distance of centroid of area A_M from left-hand support (Figure 7.5)

    $(EI)_n$ = flexural rigidity of span n

A two-hinged rectangular portal frame is once statically indeterminate, and the condition of concordancy is

$$\frac{A_{M1}\bar{x}_1}{(EI)_1 l_1} + \frac{A_{M2}}{(EI)_2} + \frac{A_{M3}(l_3-\bar{x}_3)}{(EI)_3 l_3} = 0$$

where the suffices 1 and 3 indicate the vertical legs and 2 the transom.

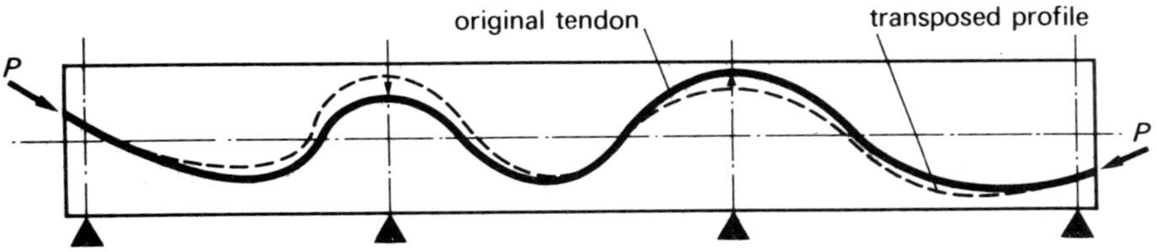

Figure 7.13 Transposed tendon in a continuous beam

A fixed rectangular portal frame prestressed by an unsymmetrical system of tendons is three times statically indeterminate, and the three conditions of concordancy are:

$$\frac{A_{M1}(l_1 - \bar{x}_1)}{(EI)_1} + \frac{A_{M3}\bar{x}_3}{(EI)_3} = 0$$

$$\frac{A_{M1}}{(EI)_1} + \frac{A_{M2}(l_2 - \bar{x}_2)}{(EI)_2 l_2} = 0$$

$$\frac{A_{M2}\bar{x}_2}{(EI)_2 l_2} + \frac{A_{M3}}{(EI)_3} = 0$$

If the tendons are arranged symmetrically the degree of indeterminacy is reduced to two and the two conditions of concordancy are

$$\frac{A_{M1}(l_1 - \bar{x}_1)}{(EI)_1 l_1} = 0$$

$$\frac{A_{M1}}{(EI)_1} + \frac{A_{M2}}{2(EI)_2} = 0$$

In a continuous prestressed beam it is possible to make a simple modification to the tendon profile without altering the resultant line of thrust. This operation is based on a property of continuous prestressed beams usually known as Guyon's Theorem, which may be defined as follows.

If the tendon profile is displaced vertically at any of the intermediate supports by any amount, but without altering its intrinsic shape between the supports, the resultant line of thrust is unchanged.

Guyon's theorem is easily proved by the tendon reaction method. Referring to Figures 7.10 and 7.13, the additional reactions introduced by the transposition of the tendon described are seen to consist of concentrated forces over each support, arising from the deflection angles in the tendons, and negligible changes in the reactions due to curvature and deflection of the tendon in the spans, arising from the small rotation of the whole tendon in each span. The former forces, acting in line with the supports cause no change of bending moment and the resultant thrust is therefore unchanged by the transposition.

The corollary of the above theorem is perhaps of greater practical importance. It will be recalled that the effect of the secondary moment is to cause a displacement of the line of thrust from the tendon profile by a vertical distance M/P at each intermediate support. It therefore follows that the tendon profile may be transposed, in accordance with Guyon's theorem, to the position of the resultant line of thrust. The resultant line of thrust therefore fulfils the conditions of concordance.

This rule is of considerable significance in the design of a tendon profile, as will be explained below.

7.6 DESIGN OF TENDON PROFILE

The method of dimensioning the sections of a statically indeterminate prestressed structure is similar to that already described for simply supported beams in Chapters 4 and 5; the maximum moment (serviceability limit state) and the minimum moment, having regard to the sign of the maximum moment, are calculated for a suitable number of points and the required

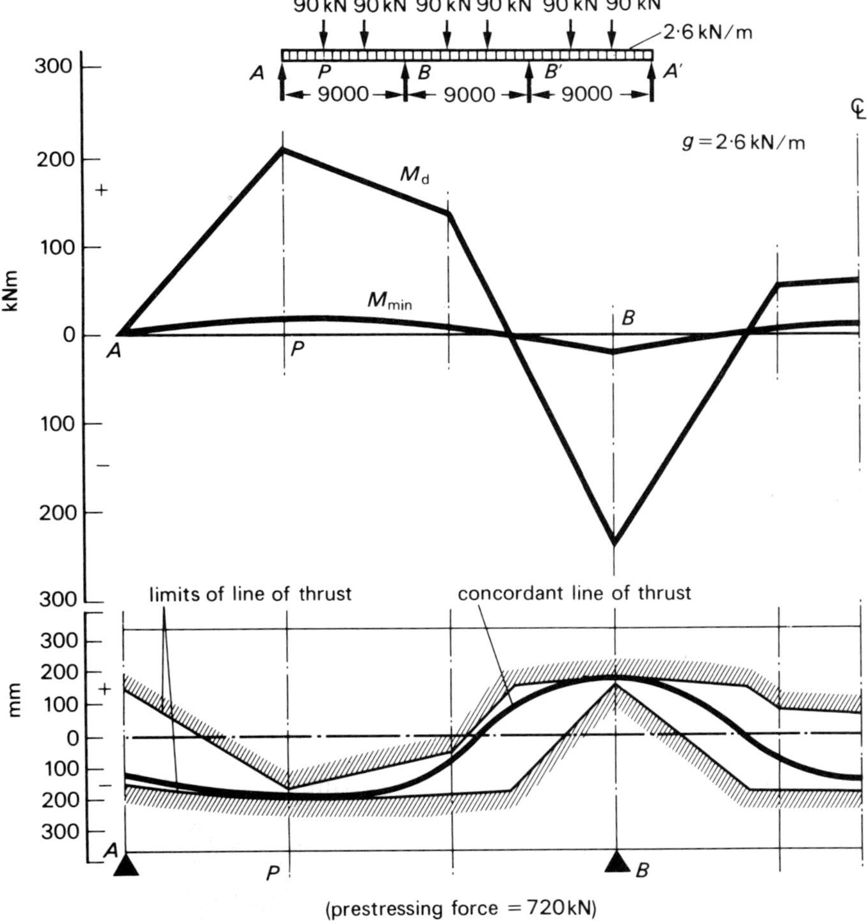

Figure 7.14 Design of tendon profile for continuous beam of three spans

section moduli are obtained from formula (4.5). The design of the tendon profile, however, is slightly more complicated, as two theoretical conditions must be fulfilled:

(1) The resultant thrust must lie within the limits defined by the allowable stresses in the concrete.

(2) The resultant thrust must satisfy the conditions of concordancy.

It will be noted that the above two conditions apply to the resultant line of thrust, rather than to the tendon profile itself. Having arrived at a suitable trajectory for the resultant thrust, this may either be adopted as the profile of a concordant tendon, or Guyon's theorem may be invoked to transpose the line of the resultant thrust to obtain a more suitable profile for the tendon.

The first of the above two conditions is the same as for a statically determinate structure, and the limits of the line of thrust are defined by the four equations (4.12) to (4.15). An example of the limits calculated for a continuous beam of three spans is given in Figure 7.14 and it may be seen that there are separate lengths of curve, those near the supports being calculated from negative moments and those in the mid-span region from positive moments. When dimensioning it is advisable to select a section of ample proportions in relation to the minimum required so that the possible positions of the line of thrust are not too severely restricted in the region of the supports and mid-span, as this may lead to difficulties in meeting the second condition, and to excessive curvature resulting in large losses of prestress due to friction.

A certain measure of trial and error is unavoidable in setting out a line of thrust that will satisfy the conditions of concordancy. Each individual design is best treated on its own merits, first fixing the eccentricity at the points where the limits do not allow much latitude and then determining the remaining dimensions so as to satisfy the appropriate concordancy equations.

For example, in the continuous beam of three spans, shown in Figure 7.14, the position of the resultant line of thrust is constrained by the close limits at P and B and the eccentricity may be taken to be fixed at -180 mm and $+190$ mm at these two points. At A the eccentricity is given a value of 125 mm so that the line of thrust lies close to the lower limit, to minimize the curvature of the tendon, and for the same reason the curved length between P and B is formed from two curves of equal length and radius, with the point of inflection close to the upper limit. The line of thrust in the member AB is thus predetermined, according to the dimensions given in Figure 7.15.

At the mid-span point of the member BB' the eccentricity of the thrust is fixed at -170 mm, close to the lower limit, leaving the length a of the curve to be calculated so that the concordancy condition is satisfied. Making use of the symmetry of the, beam, this condition is:

$$A_{\mathrm{M\,AB}}\bar{x}_{\mathrm{AB}}/l_{\mathrm{AB}} + A_{\mathrm{M\,BB'}}/2 = 0$$

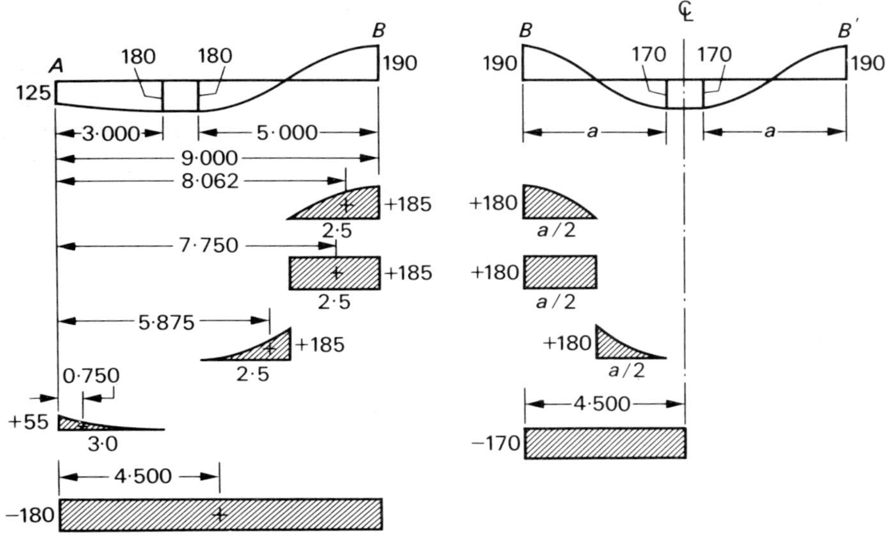

Figure 7.15 Areas for design of tendon profile for continuous beam of three spans

The area between the line of thrust and the beam axis is now divided into component areas (Figure 7.15) with the length a as the unknown quantity. Hence,

$$A_{M\,AB}\bar{x}_{AB} = P[(2/3) \times 185 \times 2\cdot5 \times 8\cdot062 + 185 \times 2\cdot5 \times 7\cdot75 + (1/3)$$
$$\times 185 \times 2\cdot5 \times 5\cdot875 + (1/3) \times 55 \times 3\cdot0 \times 0\cdot75 - 180$$
$$\times 9\cdot00 \times 4\cdot500]$$
$$= -273P$$

$$A_{M\,BB'} = P[(2/3) \times 180 \times a/2 + 180 \times a/2 + (1/3)$$
$$\times 180 \times a/2 - 170 \times 4\cdot500]$$
$$= (180a - 765)P$$

The concordancy condition may thus be written:

$$-273P/9\cdot000 + (180a - 865)P = 0$$

Therefore $a = 4\cdot420$ m

The profile thus obtained, which is shown in Figure 7.14, is suitable for the trajectory of a concordant tendon. In this case there is little point in transposing the tendon, but in some instances such a step would be advantageous, either to reduce the curvature of the tendon, or to achieve a better position of the tendon for the ultimate limit state.

7.7 AXIAL DEFORMATION AND TERTIARY MOMENTS

Throughout the preceding calculations in this chapter the effect of axial deformation has been ignored and only bending has been considered, as is the usual practice in the analysis of continuous beams and framed structures. Axial contraction, however, occurs in all prestressed members and it is now necessary to consider whether it is likely to be an important factor in the analysis of statically indeterminate prestressed structures.

Two effects have to be considered together. The first of these is the possible

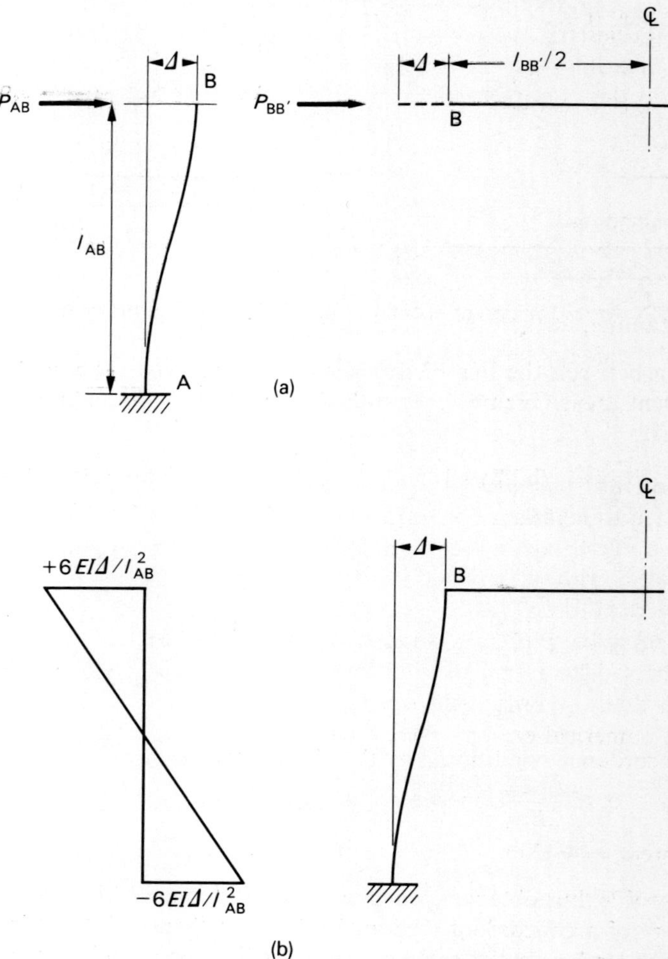

Figure 7.16 Effect of axial shortening of transom on prestressing force and secondary fixed-end moments in a fixed portal frame

reduction of the prestressing force applied along the line of a particular member by the lateral restraint of the adjacent members. The second effect is the bending deformation set up in an indeterminate structure by the contraction of the members under the action of the prestressing force. These two effects will be examined in the example of a fixed portal frame.

In this example the contraction of the vertical legs under prestress will obviously be unaffected, but referring to Figure 7.16 it will be clear that the prestressing force applied to the transom at B will be distributed, one component P_{AB} being applied transversely to the leg AB, and the remainder, $P_{BB'}$ being the effective prestressing force applied to the transom BB'.

For compatibility, Δ the lateral deflection of the end B of the leg AB must be equal to the axial contraction of one-half of the transom BB'.

Therefore,

$$\frac{P_{AB}l_{AB}{}^3}{12EI_{AB}} = \Delta = \frac{P_{BB'}l_{BB'}}{2EA_{BB'}}$$

$$\frac{P_{AB}}{P_{BB'}} = \frac{6l_{BB'}I_{AB}}{l_{AB}{}^3 A_{BB'}}$$

For example, in the example of a fixed portal used earlier in this chapter (Figure 7.7), if $A_{BB'} = 20\ 000\ \text{mm}^2$

$$\frac{P_{AB}}{P_{BB'}} = \frac{6 \times 6000 \times 208 \times 10^6}{5000^3 \times 20\ 000} = 0.003$$

For practical purposes the whole of the prestressing force may therefore be assumed to be taken by the member BB'.

If the point B moves horizontally under the action of the force P, as in Figure 7.16, without rotation of the member AB at B, an anticlockwise moment equal to $6EI_{AB}\Delta/l_{AB}{}^2$ must be applied to the member at this point, with an equal balancing at A. Treating these as fixed-end moments, the moment distribution method may be employed to calculate the tertiary moments due to the axial shortening of BB'.

In the numerical example

$$\Delta \simeq \frac{P_{BB'}l_{BB'}}{2EA_{BB'}} = \frac{140 \times 6000}{2 \times 30 \times 20\ 000} = 0.70\ \text{mm}$$

$$M_{FAB} = M_{FBA} = \frac{-6EI_{AB}\Delta}{l_{AB}{}^2} = \frac{-6 \times 30 \times 208 \times 10^6 \times 0.70}{5000^2 \times 1000}$$

$$= \doteq 1.05\ \text{kNm}$$

By moment distribution, the tertiary moment will be found to be $+0.38$ kNm (i.e. sagging) in BB' and to vary linearly in AB from -0.72 kNm at A to

+0·38 kNm at **B**. In this example the tertiary moment is thus less than 25% of the secondary moment and less than 10% of the primary moment due to the eccentricity of the tendon.

7.8 REDISTRIBUTION OF MOMENTS—THE ULTIMATE LIMIT STATE

In checking statically indeterminate prestressed structures at the ultimate limit state the flexural strength of the members at the required points is calculated by the methods described in Chapters 4 and 5. A conservative estimate of the ultimate resistance of the structure may be obtained by assuming failure to occur when the moment, calculated by the linear elastic theory, first becomes equal to the flexural strength at any section. The actual ultimate load will in most instances be somewhat greater, however, on account of a favourable redistribution of moments in the structure.

Redistribution of moments arises from the fact that the flexural rigidity *EI* of a prestressed concrete member at a given section is not constant over the full range of moment, as is assumed in the linear elastic theory, but decreases with increasing moment. A significant decrease commences with the onset of flexural cracking of the concrete and continues progressively until failure, by which stage the stiffness has been considerably reduced by cracking and by inelastic deformation of the steel and concrete. As a result of the large deformation near failure, there tends to be a concentration of curvature over a few clearly defined short lengths of member, which are sometimes represented as hinges at which a rotation is considered to take place. The upper limit of the strength of the structure, corresponding to complete redistribution of moment, is attained when the flexural strength is developed at a set of points, at which the introduction of hinges would bring about collapse by transforming the structure, or part of the structure, into a mechanism. In practice, complete redistribution does not necessarily take place, depending on the amount of hyper-elastic deformation (or rotation of the 'hinges') that can occur, and the detailed analysis of statically indeterminate prestressed structures at the ultimate limit state has been considered, both theoretically and experimentally, elsewhere, refs. (7.1, 7.2).

The designer is permitted by CP 110, 4.2.2, to allow a limited amount of redistribution of moments when checking a structure at the ultimate limit state. This is governed by four conditions:

(1) Equilibrium between the internal forces and the external loads must be maintained under each appropriate combination of ultimate loads.

(2) The ultimate resistance moment provided at any section of a member must not be less than 80% of the maximum moment at that section calculated by the linear elastic theory for all possible combinations of ultimate load.

(3) The elastic moment at any section in a member due to a particular

combination of ultimate loads must not be reduced by more than 20% of the numerically largest moment given anywhere by the elastic maximum moments diagram for that particular member, covering all appropriate combinations of ultimate load (the permissible reduction is restricted to 10% for structures over four storeys in height, in which the structural frame provides the lateral stability).

(4) Where, as the result of redistribution, the ultimate resistance moment at a section is reduced, the neutral axis depth x of the section resisting the reduced moment should not be greater than

$$x = (0{\cdot}5 - \beta_{\text{red}})d$$

where d = effective depth

β = ratio of reduction in resistance moment to the numerically largest moment given anywhere by the elastic maximum moments diagram for that particular member, covering all appropriate combinations of ultimate load.

The second of the above conditions is a safeguard against excessive cracking at the serviceability limit state. Near the points of contraflexure the moment may be reduced or even reversed as a result of redistribution as the load increases from the serviceability limit state to the ultimate limit state, and a design based solely on the latter could result in insufficient reinforcement to control cracking at the service load.

Condition (3) is a general restriction on the amount of deformation at the critical sections which accompanies redistribution and the fourth condition relates the amount of redistribution to the stiffness of the critical section, as indicated by the depth of the neutral axis.

An alternative formula, giving the permissible redistribution in terms of the reinforcement indices, will be found in ACI 318–71, 18.12.

Example
The application of these rules will be demonstrated by considering the ultimate limit state of the three-span continuous beam shown in Figure 7.14. The cross-section is given in Figure 7.17, the tensile strength of the tendon is 1020 kN, and the strength of the concrete is 40 N/mm².

Assuming that the uniformly distributed characteristic load of 2·6 kN/m is all dead load and that the concentrated characteristic loads of 90 kN each consist of 50 kN dead load and 40 kN live load applied at the four points simultaneously, the design ultimate loads are as follows:

$$\text{uniformly distributed load, } g_{\text{ud}} = 1{\cdot}4 \times 2{\cdot}6 = 3{\cdot}64 \text{ kN/m}$$

$$\text{concentrated loads, } \quad F_{\text{ud}} = 1{\cdot}4 \times 50 + 1{\cdot}6 \times 40 = 134 \text{ kN}$$

The design ultimate moments are calculated by the linear elastic theory

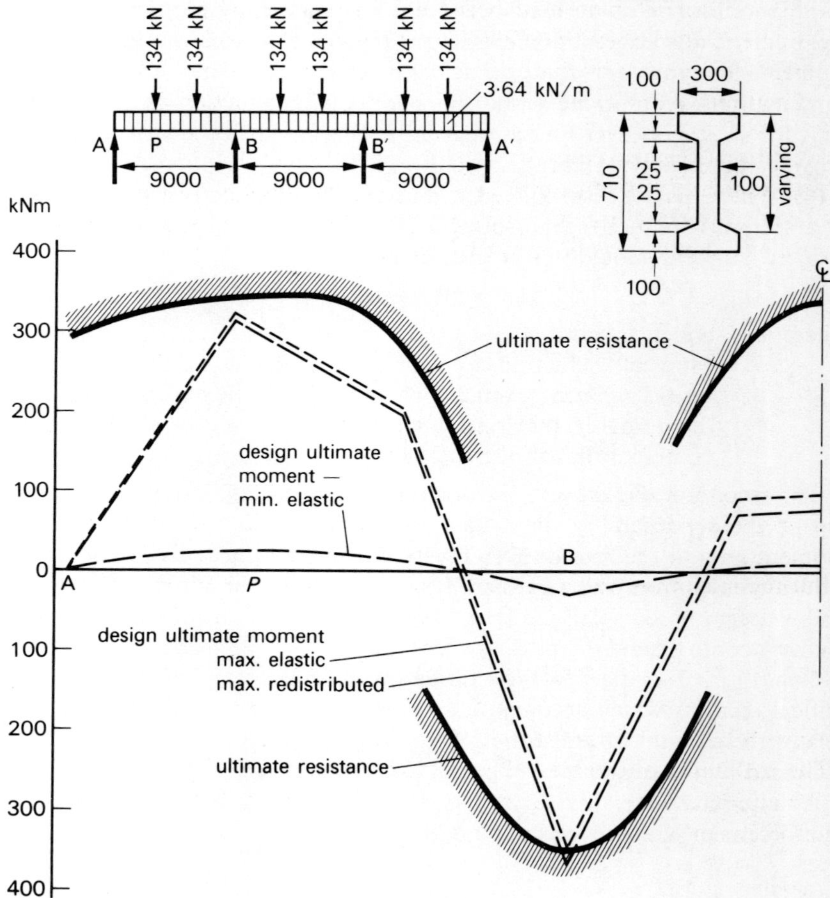

Figure 7.17 Redistribution of moments in continuous beam at ultimate limit state

and are shown in Figure 7.17. The ultimate resistance moment is calculated (see Chapters 4 and 5); this will have a positive and a negative value varying according to the position of the post-tensioned tendon, and has also been added to Figure 7.17.

At the point B, the design ultimate moment (365·8 kNm) exceeds the ultimate resistance moment (348·6 kNm). Advantage is therefore taken of moment redistribution to reduce the former by 17·2 kNm. The remainder of the diagram is drawn so that the difference between the span moment and terminal moment diagram remains equal to the 'simply supported' moment for each span, thereby fulfilling the first of the four conditions in CP 110, 4.2.2. The redistributed maximum moment is now less than the resistance at all points.

Condition (2) is not violated, as the ultimate resistance moment at B is $348\cdot6/365\cdot8 = 95\%$ of the elastic maximum moment at that point.

Condition (3) is also covered since the elastic moment at B is the numerically largest moment for the member AB.

Considering condition (4), the reduced moment at B is $348\cdot6$ kNm. The neutral axis depth x is obtained from formula (4.24)

$$M_{wu} = M_{ud} - 0\cdot2 f_{cu}(b - b_w)d_f(2d - d_f)$$

$$= 348\cdot6 - 0\cdot2 \times 40(300 - 100)112\cdot5(2 \times 545 - 112\cdot5)/10^6$$

$$= 172\cdot6 \text{ kNm}$$

$$x/d = 1 - [1 - 5M_{wu}(/f_{cu}b_w d^2)]^{\frac{1}{2}}$$

$$= 1 - [1 - 5 \times 172\cdot6 \times 10^6/(40 \times 100 \times 545^2)]^{\frac{1}{2}}$$

$$= 0\cdot477$$

But the maximum value permitted is

$$x/d = 0\cdot5 - \beta_{red}$$

$$= 0\cdot5 - (365\cdot8 - 348\cdot6)/365\cdot8$$

$$= 0\cdot453$$

The maximum allowable value of the reduction ratio is found to be

$$\beta_{red} = 0\cdot5 - x/d$$

$$= 0\cdot5 - 0\cdot477$$

$$= 0\cdot023$$

The maximum amount by which the elastic ultimate moment at B may be reduced is therefore

$$0\cdot023 \times 365\cdot8 = 8\cdot4 \text{ kNm}$$

The ultimate moments are therefore redistributed by reducing the moment at B to

$$365\cdot8 - 8\cdot4 = 357\cdot4 \text{ kNm}$$

The ultimate resistance moment at B must be increased to this value (i.e. by $357\cdot4 - 348\cdot6 = 8\cdot8$ kNm) by the addition of supplementary reinforcement. From the point of view of good detail design it would be desirable to place a nominal amount of tensile reinforcement near both the upper and the lowerface of the beam, since the tendon which provides the main tensile reinforcement is not sufficiently near the tensile face to control flexural cracking.

7.9 CONTINUITY REINFORCEMENT

A structure formed of prestressed concrete members is sometimes made continuous by the addition of reinforced concrete placed *in situ*, for example over the supports of a continuous beam or slab. This is effectively a composite indeterminate structure and must be analysed in two stages. First the prestressed members are considered in their statically determinate condition up to the stage at which the *in situ* concrete is placed. In the second stage, after the *in situ* concrete has hardened the structure has become statically indeterminate and must be analysed on this basis.

In general the *in situ* parts of the structure will be treated as ordinary reinforced concrete with regard to serviceability and for calculation of the moment redistribution at the ultimate limit state.

Although no secondary moments will be present initially, in the course of time differential shrinkage and creep will result in restraint moments at the inner supports. The shrinkage of a top slab, as discussed in Chapter 6 (p. 100ff) will cause a hogging moment equal to $\Delta\epsilon_{cs}E_cA_iy$, and it is recommended in CP 110, 5.4.3.6 that this should be multiplied by a reduction factor ϕ of 0·43 to allow for creep. Creep resulting from dead load on the beams and prestress in the precast units will further modify the restraint moment and it is recommended that the restraint moment due to prestress should be taken as the restraint moment which would have been set up if the composite section as a whole had been prestressed, multiplied by a reduction coefficient ϕ_1 of 0·87.

It is assumed above that the ratio β_{cc} of total creep to elastic deformation is equal to 2·0. For higher values of β_{cc} the reduction factors should be calculated from the expressions:

$$\phi = \frac{1 - e^{-\beta_{cc}}}{\beta_{cc}}$$

$$\phi_1 = 1 - e^{-\beta_{cc}}$$

References

7.1 INSTITUTION OF CIVIL ENGINEERS, 'Ultimate load design of concrete structures', Research Committee Report, *Proc. Instn. Civil Engrs.*, **21**, February, 1962, pp. 400–431.

7.2 BENNETT, E. W., COOKE, N. and NAUGHTON, L. P., 'Deformation of continuous prestressed concrete beams and its effect on the ultimate load', *Proc. Instn. Civil Engrs.*, **37**, May, 1967, pp. 57–74; Disc. **38**, December, 1967, pp. 769–774.

Index